Bibliografische Information der Deutschen Nationalbibliothek

Die Deutsche Nationalbibliothek verzeichnet diese Publikation in der
Deutschen Nationalbibliografie; detaillierte bibliografische Daten sind
im Internet über http://dnb.d-nb.de abrufbar.

ISBN 978-3-8325-2310-7

Logos Verlag Berlin GmbH
Comeniushof, Gubener Str. 47,
10243 Berlin
Tel.: +49 030 42 85 10 90
Fax: +49 030 42 85 10 92
INTERNET: http://www.logos-verlag.de

Regularization of inverse problems and inexact operator evaluations

von Thomas Bonesky

Dissertation

zur Erlangung des Grades eines Doktors der Naturwissenschaften
- Dr.rer.nat -

Vorgelegt im Fachbereich 3 (Mathematik & Informatik)
der Universität Bremen
im Juli 2009

Datum des Promotionskolloquiums: 11. September 2009

Gutachter: Prof. Dr. Peter Maaß (Universität Bremen)
Prof. Dr. Dirk A. Lorenz (Technische Universität Braunschweig)

Zusammenfassung

Die vorliegende Arbeit ist ein Beitrag zur Theorie der inversen Probleme mit "sparsity constraints". In den vergangenen Jahren hat sich dieser Bereich der Theorie der inversen und schlecht gestellten Probleme rasant weiterentwickelt. Es zeigte sich, dass viele inverse Probleme eine Lösung mit dünner Darstellung besitzen, d.h. dass sie mit Hilfe endlich vieler Elemente einer geeigneten Basis oder eines Frames darstellbar sind. Um solche Lösungen zu rekonstruieren, wurden in den vergangenen Jahren Tikhonov-artige Regularisierungsverfahren intensiv untersucht.

Die Minimierungsverfahren für die zugehörigen Tikhonov-Funktionale erfordern stets die Auswertung des zu Grunde liegenden Operators, sowie dessen Adjungierten. Ein Schwerpunkt dieser Arbeit ist die Untersuchung eines solchen Minimierungsverfahrens unter der Annahme, dass alle nötigen Operatorauswertungen nicht exakt berechnet werden, sondern mit Hilfe eines adaptiven Verfahrens geschehen.

Ein zweiter Hauptteil beschäftigt sich mit der Kopplung des Diskrepanzprinzips von Morozov und Tikhonov-Regularisierung, wobei der klassische quadratische Strafterm durch ein allgemeineres konvexes Funktional ersetzt wurde.

Schließlich wird ein nicht-triviales inverses Wärmeleitproblem aus der Stahlproduktion mit einer Kombination aus iteriertem Soft-Shrinkage und einem adaptiven Finite-Elemente-Verfahren gelöst.

Abstract

This thesis contributes to the field of inverse problems with sparsity constraints. In recent years this has been a rapidly developing field within the theory of inverse and ill-posed problems. It turned out that solutions of many inverse problems have a sparse structure, which means that they can be represented using only a finite number of elements of a suitable basis or frame. To reconstruct these solutions, Tikhonov-type regularization schemes have been investigated intensively within the last years.

The minimization schemes for the related Tikhonov functionals require the evaluation of the underlying operators and their adjoints. One of the main topics of this thesis is the investigation of such a minimization scheme assuming that the necessary operator evaluations are not calculated exactly, but are computed via an adaptive scheme.

A second major part is the coupling of Morozov's discrepancy principle and Tikhonov regularization, where the classical quadratic penalty term has been substituted by a more general convex functional.

Finally, a non-trivial inverse heat conduction problem from steel production is solved by a combination of iterated soft-shrinkage and an adaptive finite element method.

I would like to thank:

My supervisor Prof. Dr. Peter Maaß for advising, supporting and encouraging me over the last three years.

Dr. Kristian Bredies, Prof. Dr. Stephan Dahlke, Kamil Kazimierski, Prof. Dr. Dirk Lorenz and Prof. Dr. Thorsten Raasch for advice and fruitful discussions.

Dr. Claus-Justus Heine, Thilo Mooshagen and Bettina Suhr for their patience and help with ALBERTA.

Christina Brandt, Kylie MacDonald and Dennis Trede for discussions, reading parts of the thesis and making lots of useful remarks.

My parents Inge and Norbert who have been supporting me all my life, thank you so much.

The Deutsche Forschungsgemeinschaft for funding.

And finally all my colleagues at the Zentrum für Technomathematik for the excellent working atmosphere.

Contents

Introduction

The theory of inverse and ill-posed problems came to mathematical interest in the early sixties of the 20th century. Since that time inverse problems have been studied intensively leading to a well developed mathematical theory, especially for the Hilbert space case, see, e. g., [20, 33, 45]. Nevertheless, there are some parts of the theory, for instance inverse problems in Banach spaces or the reconstruction of sparse solutions, which have become more and more important in recent years.

Inverse problems are often modelled as operator equations

$$Ax = y,$$

with a linear or non-linear operator A mapping between some topological spaces. It is assumed that the operator A is known, as well as the data y, which are often just available as a noisy version y^δ. The solution of the inverse problem or in other words the reconstruction of the input data x is often unstable and requires regularization techniques.

Especially, regularization with sparsity constraints, i. e. the solution x has a sparse representation in some basis or frame, turned out to be a powerful tool in many applications, e. g. signal and image processing (computerized tomography, spectroscopy) or parameter identification problems (electrical impedance tomography, inverse heat conduction).

This work takes up some aspects arising from such applications and adds to the existing theory. The theoretical results presented in this thesis are motivated by an inverse heat conduction problem from the steel industry. This application will serve as a non-trivial numerical example at the end of this thesis.

A major goal of this thesis is to combine regularization with sparsity constraints with adaptive solution methods, e. g. finite element or wavelet Galerkin methods, for linear operator equations in Hilbert spaces. Especially when the evaluated operator is the solution operator of a partial differential equation, which is usually the case for parameter identification problems, this seems to be a natural setting.

The basis for the investigations in this thesis is a method which was proposed in 2004 by Daubechies, Defrise and De Mol in their pioneering paper [16] on regularization with sparsity constraints. It will be shown for a general and abstract setting that the combination of their approach with adaptive evaluation schemes is possible and forms a reliable regularization scheme.

One of the most challenging tasks in regularization theory is to develop strategies for choosing the so-called regularization parameter. Since the quality of the regularized solution crucially depends on a proper choice of the regularization parameter, it is essential to have appropriate parameter choice strategies. Therefore, the second main part of this thesis is the investigation of Morozov's discrepancy principle for Tikhonov-type functionals incorporating non-standard penalty terms. These considerations are motivated by the following facts. Most regularization results on non-standard Tikhonov regularization presume an a priori choice of the regularization parameter. This approach requires knowledge on the smoothness of the unknown solution and provides us only with an order of magnitude for an appropriate choice of the regularization parameter. However, in practice this smoothness information may not be available. It may be more suitable to choose the regularization parameter in such a way that the discrepancy $\|Ax_\alpha^\delta - y^\delta\|$, including the regularized solution x_α^δ, is of the same order as the noise level δ of the given data. This leads to a posteriori parameter choice rules, taking the regularized solution and the noisy data into account. Hence, as a commonly used method Morozov's discrepancy principle is investigated.

As mentioned above regularization with sparsity constraints is a field within the theory of inverse problems which has evolved during the last few years. Since the initial article [16] has been published, there has been an enormous amount of publications refining and extending the basic ideas of sparsity reconstruction. The fundamental idea of this theory is to consider the minimization of Tikhonov-type functionals

$$\Gamma_{\alpha,\delta}(x) = \|Ax - y^\delta\|^2 + \alpha\Omega(x), \tag{1}$$

where the classical penalty term $\Omega(x) = \|x\|_H^2$ including the norm of the underlying Hilbert space H is substituted by the power of a weighted sequence norm

$$\Omega(x) = \sum_\lambda w_\lambda |\langle x, \varphi_\lambda \rangle|^p = \sum_\lambda w_\lambda |x_\lambda|^p \tag{2}$$

with positive weights $w_\lambda \geq \omega > 0$, $1 \leq p \leq 2$ and an orthonormal basis $\{\varphi_\lambda\}_\lambda$ of H. If it is a priori known that the solution of the inverse problem has a sparse structure with respect to the considered orthonormal basis, i. e. the basis expansion of the solution incorporates just a finite number of basis elements, then this is a reasonable approach. Considering $p < 2$, the special structure of the penalty term promotes sparsity of the minimizer, i. e. coefficients x_λ with $|x_\lambda| < 1$ are more penalized than in case of $p = 2$. This effect is enhanced the smaller p is chosen.

The iterative minimization scheme for such Tikhonov-type functionals proposed in [16] consists of a Landweber step followed by a non-linear shrinkage procedure

$$x^{n+1} = \mathbf{S}_{\alpha\mathbf{w},p}(x^n - A^*(Ax^n - y^\delta))$$

and is called iterated soft-shrinkage. In case of $p = 1$ it will also be called iterated soft-thresholding in the upcoming chapters. Further results related to this iteration can be found, e. g., in [8, 9], where the equivalence to a generalized gradient method and convergence properties were analyzed. An extension to non-linear inverse problems was investigated, e. g., in [4, 40, 41]. It turns out that, in case of non-linear operators often an additional fixed point iteration is needed to determine the minimizers of the Tikhonov-type functionals. Further, in the non-linear case the minimizer is not unique in general. Hence, only convergence to a local minimizer can be shown.

Apart from the original algorithm, many other minimization schemes, especially for the case where $p = 1$, were developed and some of them should be mentioned in this context. In [15] the authors consider the minimization of the sum of two convex functionals via forward-backward splitting which covers also the case of sparsity promoting Tikhonov-type functionals. The article [22] deals with the case of vector valued data where the components show a joint sparsity pattern. In [17] the connection between a combination of Landweber iteration and ℓ_2-projections onto ℓ_1-balls and iterated soft-shrinkage is investigated. The reference [21] deals with domain decomposition methods and iterated soft-shrinkage, the authors in [7] modify the Tikhonov-type functional in such a way that the minimizer stays the same, but the iterated soft-shrinkage iteration changes into a hard-shrinkage iteration and finally the paper [25] is mentioned, where a semi-smooth Newton method is investigated to calculate the minimizers.

Currently Tikhonov-regularization with non-convex penalty terms is a growing field of research. Especially for penalty terms (2) with $p < 1$ some first results have been released recently, see, e. g., [23, 32, 53].

The above listings are not complete by far, but it shows the variety of approaches which have been studied during the last few years.

All these theoretical results have one assumption in common. It is assumed that the operator, defining the inverse problem, can be evaluated exactly. However, in practice the operator always has to be discretized or approximated to do numerical calculations. In case of classical Tikhonov regularization, i. e. with $\Omega(\cdot) = \| \cdot \|^2$, there are already some results taking inexact operator evaluations into account. It is usually assumed that the approximated operator A_h satisfies an estimate of the form

$$\|A_h - A\| \leq h. \tag{3}$$

This estimate is formulated in operator norm and is independent of the evaluation point x. Some early remarks on Tikhonov regularization with operator approximations were made in [51]. Further articles dealing with finite dimensional approximations of the operator and considered function spaces have been published by Neubauer [37] and Neubauer and Scherzer [38]. Whereas the first paper deals with linear operators, the latter handles non-linear ones. In both cases optimal convergence rates were also proved. A more general result for Tikhonov regularization in combination with a discrepancy principle and operator approximations has been published by Maass and Rieder [34]. Again, optimal convergence rates were proved in this case.

All results listed so far are dealing with classical Tikhonov regularization. An article which goes more into the direction of sparsity reconstruction was written by Combettes and Wajs [15]. The authors investigated an iterative minimization scheme for a sum of two convex functionals. The Tikhonov-type functional (1) fits into their framework. The considered iterative minimization scheme contains additive terms which can be interpreted as error terms caused by operator approximations. To be able to prove convergence, the authors assumed summable errors.

Neither the assumption (3) nor the assumption of summable errors seems to be suitable in the case of adaptive operator evaluations, especially when the operator is the solution operator of a partial differential equation and adaptive schemes are used for evaluation. In this case, it is more appropriate to assume a pointwise error estimate like

$$\|[Ax]_h - Ax\| \leq h$$

for all $x \in H$, with approximate solution $[Ax]_h$, where the realization of the adaptive scheme $[A\cdot]_h$ may vary in every point x. In this thesis an adaptive approach and the original soft-shrinkage algorithm presented in [16] are combined, which leads to an iteration of the form

$$x^{n+1} = \mathbf{S}_{\alpha\mathbf{w},p}(x^n - [A^*([Ax^n]_h - y^\delta)]_{h^*}).$$

We show that this iteration produces iterates, which will be located in some ε-ball around the minimizer of the corresponding Tikhonov functional after a finite number $N = N(h)$ of iteration steps. The radius ε of these balls is coupled to the error level h and gets arbitrarily small as the error level tends to zero. Further, we will prove regularizing properties and convergence rates. In particular, this means that we will investigate how the error level h, the regularization parameter α and the noise level δ of the data have to be coupled to form a regularization method.

A recently developed method, which goes in the same direction as the approach presented in this thesis, has been published by Ramlau, Teschke and Zhariy [42]. They make use of three routines introduced by Stevenson in [50] to discretize and solve the normal equation $A^*Ax = A^*y$ corresponding to the considered operator equation $Ax = y$. They reach at a Landweber iteration of the form

$$x^{n+1} = \mathbf{COARSE}[x^n - \beta\mathbf{APPLY}[x^n] + \beta\mathbf{RHS}[y^\delta]],$$

where the three subroutines depend on several parameters which is not indicated at this point. The operator $\mathbf{RHS}[\cdot]$ creates an approximation of A^*y^δ which satisfies some error tolerance and can be represented by finitely many expansion coefficients with respect to some preassigned system of functions. The routine $\mathbf{APPLY}[\cdot]$ performs the application of the operator A^*A to x^n and returns again a finitely represented approximation of the exact solution. Finally, the $\mathbf{COARSE}[\cdot]$ operator returns an approximate version of the given argument which is determined by the most significant expansion coefficients of the argument and fulfills again some error estimates. This coarsening procedure acts in some sense like the shrinkage operator in the approach presented in this thesis and also leads to a sparse solution with respect to the preassigned system of functions.

It is interesting that this iteration creates sparse solutions and looks quite similar to our approach, although it is not based on the minimization of Tikhonov-type functionals. Comparing the iterative schemes, the main differences to our method are the different discretization and the use of the coarsening routine instead of the soft-shrinkage operator. Another important difference is the regularization parameter. Whereas the regularization parameter in our approach balances the two terms of the underlying Tikhonov-type functional and has to be coupled carefully to the noise level of the data and error tolerance of the adaptive evaluation scheme, the regularization parameter for the Landweber iteration is the iteration index itself, which has to be coupled to the noise level and the three error tolerances of the adaptive subroutines.

Thinking of solution operators of PDEs it seems as if our approach would be more natural, since the underlying PDEs are solved with some adaptive scheme which can be used as a black box. Everything which has to be known is the error compared to the true solution. The other approach requires a specific way of discretizing the problem as well as using the routines $\mathbf{RHS}[\cdot]$, $\mathbf{APPLY}[\cdot]$ and $\mathbf{COARSE}[\cdot]$.

The quality of solutions of inverse problems crucially depends on the choice of the regularization parameter. There are various parameter choice strategies, most of which can be separated into a priori and a posteriori strategies. Whereas an a priori parameter choice only takes the noise level δ of the data into account, an a posteriori strategy also depends on the regularized solution x_α^δ. Further, a priori strategies usually need information on the smoothness of the unknown solution of the inverse problem to determine the best or in some sense optimal choice of the regularization parameter.

Considering classical Tikhonov regularization, there are results on the optimal parameter choice with respect to some operator dependent Hilbert scales, see, e. g., [20, 33] for details. It is assumed that the unknown solution $x^\dagger$ is located in some function space $X_\nu = R((A^*A)^{\nu/2})$ or equivalently that $x^\dagger$ fulfills a so-called source condition:

$$\text{a } z \text{ exists such that } x^\dagger = (A^*A)^{\nu/2}(z). \tag{4}$$

With such an assumption, optimal convergence rates can be proved for classical Tikhonov regularization.

In case of Tikhonov-type functionals, i. e. where the penalty term has been changed into some convex functional, some results have been developed in recent years, trying to generalize the classical results. Some of the latest work on convergence rates for Tikhonov-type regularization methods can be found in [11, 28, 44]. In each case the source condition (4) has been generalized and an a priori choice of the regularization parameter has been taken into account. These results provide us only with an order of magnitude, but not with a concrete value for the regularization parameter. All results assume a parameter choice like $\alpha \sim \delta^\sigma$, which ensures a theoretical convergence rate of the regularization scheme. However, to determine a regularized solution we need a concrete value of α. Since the regularized solutions may change significantly while varying the regularization parameter in a small range, it may be better to follow other parameter choice strategies.

One possibility is to use discrepancy principles. The well-known discrepancy principle by Morozov [36], see also [20, 33], is based on the solution of the non-linear equation

$$\|Ax_\alpha^\delta - y^\delta\| = \delta. \tag{5}$$

Since this equation is in general highly non-linear with respect to α, it is usually solved approximately, choosing α in such a way that

$$\|Ax_\alpha^\delta - y^\delta\| \le \tau\delta$$

is satisfied, with some $\tau > 1$. This approach means that α is chosen such that the discrepancy on the left hand side is of the same order as the noise level of the data, which is the smallest discrepancy we should ask for. Otherwise the quality of the solution would be better than the data. An advantage of this approach is that the parameter choice can be done in an iterative way. We pick some α_0, compute $x_{\alpha_0}^\delta$ and decrease the parameter until the discrepancy falls below the bound $\tau\delta$.

Since in this thesis the focus of our attention is on sparsity reconstruction, Morozov's discrepancy principle is combined with Tikhonov-type functionals including the case of functionals

$$\Gamma_{\alpha,\delta}(x) = \|Ax - y^\delta\|^2 + \alpha \sum_\lambda w_k|x_\lambda|^p$$

as described above. Nevertheless, we consider a more general class of basically convex, weakly coercive, positive and lower semi-continuous penalty terms Ω. Assuming some additional properties, we will be able to prove that the combination with Morozov's discrepancy principle provides us with a regularization method. In addition, a general convergence rate result in terms of Bregman distances can be shown, if an appropriate source condition, i. e.

$$\text{a } z \text{ exists, such that } A^* z \in \partial \Omega(x^\dagger),$$

is assumed. One possibility for proving convergence rate results in terms of the norm of the underlying Hilbert space H, is assuming q-convexity of the functional defining the penalty term Ω. It turns out that in case of sparsity promoting penalty terms introduced by Daubechies et al., this assumption is fulfilled, which results in a convergence rate of $\mathcal{O}(\delta^{1/2})$. Further, in case of sparsely represented solutions this rate can be improved if the operator satisfies the so-called finite basis injectivity property. This basically means that the operator A restricted to every subspace of H spanned by finitely many basis elements of a given orthonormal basis is injective. In this case the convergence rates can be even improved to $\mathcal{O}(\delta^{1/p})$.

At this point it has to be remarked that further results have been worked out combining Morozov's discrepancy principle and Tikhonov-type functionals. They can be found in [30]. The authors started from a more general framework considering operator equations in Banach spaces and more general discrepancy terms. They investigated an iterative method to solve a non-linear equation corresponding to (5) directly.

Finally, the organization of the upcoming parts of this thesis is as follows:

In the first chapter a survey on the theory of inverse and ill-posed problems is given. Further important results from the theories of convex analysis and geometry of Banach spaces are introduced, which will be needed later on to investigate Tikhonov-type functionals. The last part of Chapter 1 is dedicated to sparsity reconstruction. Some basic facts on iterated soft-shrinkage, regularizing properties and convergence rates of the Tikhonov-type regularization method proposed by Daubechies et al. are discussed.

In Chapter 2 regularization with sparsity constraints taking inexact operator evaluations into account is considered. This chapter is divided into three sections. The first one deals with iterated soft-shrinkage $(1 < p \leq 2)$ combined with adaptive operator evaluations. Despite the first section, in section two an estimate in operator norm, i. e. $\|A_h - A\| \leq h$, is assumed for the operator approximation A_h. Within the third section adaptive iterated soft-thresholding $(p=1)$ is investigated. Most results presented in the first section of Chapter 2 have already been published in a slightly different form in [6].

The third chapter is dedicated to Morozov's discrepancy principle. It is shown that the combination with Tikhonov-type functionals incorporating non-standard penalty terms forms a regularization scheme. Further, some results on convergence rates are presented. At the end of the chapter a first result for Morozov's discrepancy principle and iterated soft-shrinkage including adaptive operator evaluations is given. Some results of Chapter 3 have been published in [3].

The fourth Chapter deals with numerical experiments applying the theoretical work presented in the previous chapters. As a non-trivial example an inverse heat conduction problem from steel industry is investigated. The solution is calculated with a combination of an adaptive finite element solver based on the toolbox ALBERTA2, see [46] and the iterated soft-shrinkage algorithm.

In the final chapter the main results are briefly summarized and some additional remarks as well as open questions are indicated.

Chapter 1

Concepts

This chapter deals with concepts and facts which will be needed in the forthcoming parts of this thesis. It begins with a brief introduction to the theory of inverse and ill-posed problems to get familiar with important definitions and statements. Further terms and results from convex analysis and the geometry of Banach spaces are introduced. The last part of the current chapter is on regularization with sparsity constraints. A survey on the existing theory is presented and some important results we will make use of in the following sections are emphasized.

1.1 Inverse problems and ill-posedness

First it has to be explained what is meant by an inverse problem and when it is called ill-posed.

Many technical or physical processes can be modeled mathematically as an operator equation of the form

$$Ax = y, \tag{1.1}$$

where A represents an operator mapping between topological spaces. In our case we restrict ourselves to Hilbert or Banach spaces. The operator acts on some input data x and represents the mathematical model of the process under consideration. The right hand side y represents the output data of the process.

If we are provided with the operator and the input data we call the problem direct. In case that the operator and the output data are available we call the problem inverse. Solving the inverse problem means to reconstruct an approximation to the input data x from the output or measurement data y. This may cause some trouble, since the inverse of the operator A may not exist or be discontinuous.

In this case, we call the inverse problem ill-posed according to the definition by Hadamard, see, e. g., [33].

Definition 1.1.1 (Hadamard) *[33, Definition 1.1.1] Let $A : X \to Y$ be a mapping between topological spaces. The problem (A, X, Y) is called well-posed if*

- *$Ax = y$ has a solution for every $y \in Y$,*

- *this solution is unique,*

1

- the solution depends continuously on the data y.

In the case that one of the conditions is not satisfied, the problem is called ill-posed.

The existence and uniqueness of solutions can be assured by defining the so-called generalized solution. This concept will be introduced briefly in the following, see, e. g., [33] for details.

Instead of solving the operator equation (1.1) directly, we minimize the discrepancy

$$\|Ax - y\|.$$

In the case of a linear operator A, mapping between Hilbert spaces X and Y, this is possible for $y \in R(A) \oplus R(A)^\perp$. In general the minimizer is not unique. To define a unique solution, we consider the minimizer with minimal norm and denote it by $x^\dagger$. This element is called the generalized solution of (1.1). The mapping

$$A^\dagger : D(A^\dagger) = R(A) \oplus R(A)^\perp \subset Y \to X$$

is called generalized inverse or pseudo-inverse. The generalized solution can also be calculated as solution of the normal equation

$$A^*Ax = A^*y$$

in $\overline{R(A^*)}$.

This solution concept ensures the existence of a unique solution for every $y \in D(A^\dagger) \subset Y$. However in most cases the generalized inverse is discontinuous. It can be shown that the generalized inverse is continuous if and only if the range of the operator A is closed. For non-degenerate operators this in turn is only the case in finite dimensional spaces.

Usually, the exact data y is not available, but just a noisy version y^δ, which satisfies $\|y - y^\delta\| \leq \delta$. Moreover, $y^\delta \in Y$ may not lie within the domain of definition $D(A^\dagger)$ of the generalized inverse. So we need strategies to overcome the problem of discontinuity of $A^\dagger$ as well as the limitation due to the domain of definition $D(A^\dagger)$.

In the upcoming section some strategies to stabilize the solution process of such inverse problems will be presented.

1.2   Regularizing inverse problems

In this section some strategies to remedy the discontinuity of the generalized inverse are presented, or generally speaking, strategies to handle the instabilities arising if we try to solve inverse and ill-posed problems. First a general definition of a regularization of the generalized inverse $A^\dagger$ is presented. Afterwards, we take a look at some special regularization methods where we focus especially on Tikhonov regularization, since this concept is essential throughout this thesis. We consider different parameter choice strategies and investigate the regularizing properties of the described schemes. Finally, we turn to Tikhonov regularization incorporating more general penalty terms than the classical squared norm.

Definition 1.2.1 *[33, Definition 3.3.1] A regularization of $A^\dagger$ is a family of continuous operators $\{T_\alpha\}_{\alpha>0}$ with*

$$T_\alpha : Y \to X$$

which has the following property:
a mapping $\alpha : \mathbb{R}^+ \times Y \to \mathbb{R}^+$ exists such that for all $y \in D(A^\dagger)$ and all $y^\delta \in Y$ with $\|y - y^\delta\| \leq \delta$ holds

$$\lim_{\substack{\delta \to 0 \\ y^\delta \to y}} T_{\alpha(\delta,y^\delta)}y^\delta = A^\dagger y. \tag{1.2}$$

In case that all T_α are linear, the family $\{T_\alpha\}_{\alpha>0}$ is called a linear regularization. And α is called the regularization parameter, which is chosen such that

$$\lim_{\substack{\delta \to 0 \\ y^\delta \to y}} \alpha(\delta, y^\delta) = 0.$$

If the choice of α does not depend on y^δ the parameter choice is called a priori, otherwise it is called a posteriori.

At first sight, this definition provides us with a quite abstract strategy to define reasonable approximations to the generalized solution $x^\dagger$. Hence, some remarks are given to emphasize important facts related to this definition.

Remark 1.2.2 *As a regularization method we understand the combination of a parameter choice rule and a family of regularizing operators.*
We call $x_\alpha^\delta = T_\alpha(y^\delta)$ a regularized solution of (1.1) considering noisy data y^δ.
Usually, we consider parameter choice rules which only depend on the noise level, since we assume that the noisy data y^δ satisfies the condition $\|y - y^\delta\| \leq \delta$. Therefore we just indicate the dependence on δ in the following, i. e. $\alpha = \alpha(\delta)$.
The most challenging task is to choose the regularization parameter α in a proper way. Especially, we have to choose α in such a way that (1.2) holds. This means

$$\|x_\alpha^\delta - x^\dagger\| \xrightarrow{\delta \to 0} 0, \tag{1.3}$$

which is referred to as the regularizing property of a regularization scheme. Moreover, α should be chosen in a somewhat optimal way, which means that the rate of convergence in (1.3), with respect to the noise level δ, is as good as possible.

To illustrate the concept of regularization, there are three commonly used regularization schemes presented. First we have a short look at the truncated singular value decomposition (TSVD), second we investigate the Landweber iteration and finally we turn to Tikhonov-type methods.

As we will see in the following, compact operators provide us with inverse and ill-posed problems. Their investigation is very helpful to understand the difficulties which arise when we try to solve inverse problems. First, to investigate the TSVD, some properties of compact operators are mentioned.

Assume $A \in K(X, Y)$ to be a non-degenerate operator, i. e. with an infinite dimensional range, mapping between separable Hilbert spaces X and Y. Then it holds that $A^*A \in K(X)$ is also non-degenerate and in addition to it self-adjoint. Especially, it has a system of eigenvectors v_n and eigenvalues λ_n with a single accumulation point at zero. We can define a singular system for A as follows.

Definition 1.2.3 *Let $A \in K(X, Y)$ be a non-degenerate operator mapping between separable Hilbert spaces. Denote by λ_n the eigenvalues of A^*A with $\lambda_1 \geq \lambda_2 \geq \cdots$ and by v_n the associated eigenvectors. The system $\{v_n, u_n, \sigma_n\}_{n>0}$ with*

$$\sigma_n = +\sqrt{\lambda_n} \quad \text{and} \quad u_n = \sigma_n^{-1} A v_n$$

is called a singular system of A.

With the help of a singular system, we can represent the operator A and its generalized inverse $A^\dagger$ in the following way

$$Ax = \sum_n \sigma_n \langle x, v_n \rangle u_n$$

and

$$A^\dagger y = \sum_{\sigma_n > 0} \sigma_n^{-1} \langle y, u_n \rangle v_n,$$

respectively, see, e. g., [33].

Since $\sigma_n \to 0$ and therefore $\sigma_n^{-1} \to \infty$, we see that $A^\dagger$ cannot be continuous. This means that regularization is necessary to stabilize the inversion process. The simplest way would be to cut off the expansion for $x^\dagger = A^\dagger y$ at some finite index. This approach is called TSVD where the truncation index acts as regularization parameter. This scheme forms a regularization method, see [33, Satz 4.1.2] for a proof.

Remark 1.2.4 *There is a general theory on regularization with so called filter functions F_α acting on the singular values, which produce regularizing operators of the form*

$$T_\alpha y^\delta = \sum_{\sigma_n > 0} F_\alpha(\sigma_n) \sigma_n^{-1} \langle y^\delta, u_n \rangle v_n.$$

In case of the TSVD the filter function is given by

$$F_\alpha(\sigma) = \left\{ \begin{array}{lll} 1 & : & \sigma \geq \alpha \\ 0 & : & \sigma < \alpha \end{array} \right. .$$

These filter functions cause a damping of the inverse singular values and have to be defined in such a way that the norm of the operators T_α remains finite for every fixed α. Besides the TSVD, the following concepts of Landweber iteration and Tikhonov regularization could also be formulated in terms of filter functions.

To present all the details on regularization with filter functions would go beyond the scope of this thesis. Therefore the interested reader is referred to [33].

If the singular system of an operator is known, the TSVD may be an easy way to stabilize the inverse problem. However, in most cases the singular system is actually not available. In the following, we will discuss regularization strategies based on minimization problems. As we saw in Section 1.1, instead of solving the operator equation (1.1) directly, we can also try to minimize the discrepancy, i. e. we solve

$$\min_{x \in X} \|Ax - y^\delta\| \quad \text{or equivalently} \quad \min_{x \in X} \Phi(x) = \min_{x \in X} \|Ax - y^\delta\|^2.$$

An iterative algorithm to minimize the functional Φ goes back to Landweber and looks like

$$x^{n+1} = x^n - \mu \nabla \Phi(x^n),$$

with some appropriate stepsize $\mu > 0$ and an arbitrary starting point x^0. This method is also called a gradient method, since the descent direction is given by the negative gradient of the functional Φ. In our case the algorithm looks like

$$x^{n+1} = x^n - \mu A^*(Ax^n - y^\delta).$$

It can be shown that it forms a linear regularization method if we set $\alpha = \frac{1}{n}$ and choose $0 < \mu < \frac{2}{\|A\|^2}$, see, e. g., [33, Satz 4.3.3].

Using Landweber iteration we produce regularized solutions by stopping the iteration at a finite index. Otherwise, the norm of the regularized solutions, the iterates, would blow up indefinitely.

In many cases it is a priori known that the norm of the ideal solution is bounded by some constant value, i. e. $\|x^\dagger\|_X \le \rho < \infty$ or some other prior information like $\Omega(x^\dagger) \le \rho < \infty$ is available, where Ω represents some functional on X. The first approach was investigated by Tikhonov, see, e. g., [51]. He proposed to minimize functionals of the form

$$\Gamma_{\alpha,\delta}(x) = \|Ax - y^\delta\|^2 + \alpha\|x\|^2, \tag{1.4}$$

where the first part, called discrepancy term, enforces the approximation of the data and the second part, called penalty term, prevents the norm of the solution from blowing up indefinitely. The regularization parameter α balances these two terms. Since the Tikhonov functionals (1.4) are strictly convex, they have a unique minimizer, which we denote by x_α^δ. We define the regularizing operators by

$$y^\delta \longmapsto T_\alpha(y^\delta) = x_\alpha^\delta = \operatorname*{argmin}_{x \subset X} \Gamma_{\alpha,\delta}(x).$$

The following theorem shows that Tikhonov's approach combined with a proper parameter choice strategy forms a regularization method. See [20] for the proof.

Theorem 1.2.5 *[20, Theorem 5.2] Let x_α^δ denote the unique minimizer of (1.4), $y \in R(A)$ and $\|y - y^\delta\| \le \delta$. If $\alpha = \alpha(\delta)$ is chosen such that*

$$\lim_{\delta \to 0} \alpha = 0 \quad \text{and} \quad \lim_{\delta \to 0} \frac{\delta^2}{\alpha} = 0, \tag{1.5}$$

then

$$\lim_{\delta \to 0} \|x_\alpha^\delta - x^\dagger\| = 0.$$

The simplest a priori parameter choice rule, which fulfills condition (1.5), is setting $\alpha \sim \delta$. It does not incorporate any information on the minimizer x_α^δ and depends only on the noise level.

As mentioned above, besides this convergence itself, how fast the regularized solutions x_α^δ approach the generalized solution $x^\dagger$ is an important question. If we think of a well posed problem, with a continuous inverse, the convergence rate should be of order δ. In the ill-posed case we expect a convergence rate which is worse than $\mathcal{O}(\delta)$.

To prove convergence rates for regularization methods we need additional information on the regularity of the unknown solution. In the classical theory it is often assumed that the solution $x^\dagger$ is located in some function space X_ν depending on the operator A, defined as

$$X_\nu = R((A^*A)^{\nu/2}).$$

Such an assumption forms a so-called source condition. There might be different formulations of source conditions, as we will see later. If we think of classical Tikhonov regularization it can be shown that the assumption $x^\dagger \in X_\nu$, with $0 < \nu \le 2$ and a parameter choice $\alpha \sim \delta^{2/\nu+1}$ leads to a convergence rate of $\mathcal{O}(\delta^{\nu/\nu+1})$. Further, it turns out that this is the best rate we can achieve. Therefore the parameter choice is called optimal, see, e. g., [33]. Hence, the best possible convergence rate for classical Tikhonov regularization is $\mathcal{O}(\delta^{2/3})$, see, e. g., [20]. This shows that the convergence rate depends essentially on the parameter choice, which is connected with the prior information on the solution.

So far we have only considered a priori parameter choices. Since the smoothness of the solution is unknown in general and the parameter ν is also not available, we cannot choose the regularization parameter according to the optimal a priori strategy. Further, such an approach provides us only with an order of magnitude for the regularization parameter depending on the noise level and the smoothness parameter ν, but not with an explicit value for α. Whereas the asymptotic behavior does not change when multiplying α by a constant number, the regularized solution may change completely.

To give an alternative, we consider an a posteriori parameter choice rule, namely the discrepancy principle introduced by Morozov [36] and investigated in more detail, e. g. in [20, 33]. The advantage of such a parameter choice is that it provides us with a concrete scheme for determining the best α. The basic idea is to compare the residual $\|Ax_\alpha^\delta - y^\delta\|$ and the noise level δ. Operating with noisy data y^δ fulfilling $\|y - y^\delta\| \le \delta$, the smallest residual we should ask for is

$$\|Ax_\alpha^\delta - y^\delta\| = \delta.$$

In practice this non-linear equation for determining the regularization parameter α is often not solved directly. Frequently the parameter is determined iteratively by choosing a decreasing sequence $\{\alpha_n\}_n$, e. g. $\alpha_n = q^n\alpha_0$, with $0 < q < 1$ and calculating the regularized solutions $x_{\alpha_n}^\delta$ until the residual fulfills

$$\|Ax_{\alpha_n}^\delta - y^\delta\| \le \tau\delta$$

for some $\tau > 1$.

Also for the combination of Morozov's discrepancy principle and classical Tikhonov regularization an optimal convergence rate can be shown. Assuming $x^\dagger \in X_\nu$, with $0 < \nu \leq 1$ leads again to an optimal rate of $\mathcal{O}(\delta^{\nu/\nu+1})$ and therefore the best convergence rate is $\mathcal{O}(\delta^{1/2})$, see, e. g., [20] for details.

Up to now, we have considered the classical approach of Tikhonov presented in (1.4). This approach takes prior information with respect to the norm of the underlying Hilbert space into account. The penalty term makes sure that the norm of the regularized solution does not blow up indefinitely. Moreover, solutions created using the classical Tikhonov functional are comparatively smooth. This can be explained by the following line of argument. As mentioned in Remark 1.2.4 Tikhonov regularization can be formulated via filter functions. This leads to the following representation:

$$T_\alpha y^\delta = \sum_{\sigma_n > 0} \frac{\sigma_n^2}{\sigma_n^2 + \alpha} \sigma_n^{-1} \left\langle y^\delta, u_n \right\rangle v_n.$$

It turns out that in case of operators defining ill-posed problems, small singular values correspond to high oszillating singular functions, see, e. g., [33]. This means that the damping of small singular values weakens the influence of high oszillating components and leads to smooth solutions.

In many situations the structure of the solution is known, e. g. when thinking of image processing, we might be provided with a noisy image and would like to deblur it. In such cases we might be interested in reconstructing the image while preserving sharp edges in the image. This can also be done with Tikhonov regularization, but replacing the squared Hilbert space norm in the penalty term by a TV-norm. Another important field in regularization theory is regularization with sparsity constraints. In this case it is assumed that the solution has a sparse re- presentation in some basis or frame of the underlying Hilbert space. To reconstruct such sparse functions, the classical penalty term is replaced by some power of a weighted ℓ^p-norm. This approach will be investigated intensively within the last section of this chapter and is a central concept throughout this thesis.

These are only two examples, which justify the extension of Tikhonov regularization theory beyond the classical approach. In what follows we investigate Tikhonov-type functionals of the form

$$\Gamma_{\alpha,\delta}(x) = \|Ax - y^\delta\|^2 + \alpha\Omega(x), \tag{1.6}$$

where we will specify Ω when it will be needed. Using this approach, we assume some prior information on the solution represented by some appropriate choice of Ω.

To deal with the described generalized Tikhonov approach, the concept of Ω-minimizing solutions has to be introduced.

Definition 1.2.6 *$x^\dagger$ is called an Ω-minimizing solution of $Ax = y$ if*

$$x^\dagger = \operatorname*{argmin}_{Ax=y} \Omega(x).$$

In case of $\Omega(\cdot) = \|\cdot\|^2$, this definition coincides with the classical generalized solution. In the following, dealing with Tikhonov regularization always means that we have Ω-minimizing solutions as ideal or true solutions in mind.

In recent years many results have been published for various variants of Tikhonov regularization, e. g. incorporating different penalty terms or considering operators mapping between Banach spaces, dealing with both minimization algorithms and regularization results. Some of the latest results for a general framework are, e. g., [11, 28, 43, 44, 48]. Further, there are various results considering sparsity reconstruction as already mentioned in the introduction. In the following, we will come back to some of these references to have a closer look at them.

1.3   Convex analysis

This thesis deals with the minimization of Tikhonov-type functionals as defined in (1.6). The properties of those functionals, especially concerning existence and uniqueness of minimizers, depend on the properties of the considered penalty terms Ω. In this context convexity of the functionals is a key property. In the following, a survey on concepts and results from the theory of convex analysis, or more specifically convex optimization, is presented. The focus is on those parts of the theory which will be needed for the special situations in the upcoming chapters of this thesis. Since the presented theory is applied to special situations, the concepts and results may not be presented in full generality. Furthermore, some of the presented concepts are not restricted to convex functionals. For results in full generality the reader is referred, e. g. to [19, 54].

At the beginning of a section on convex analysis, what is meant by a convex set and a convex functional, must be defined.

Definition 1.3.1 *Let X be a real Banach space and $f : X \to \mathbb{R} \cup \{-\infty, \infty\}$.*

- *A subset $C \subseteq X$ is called convex, if $x, y \in C$ and $t \in [0,1]$ imply $tx + (1-t)y \in C$.*

- *The functional f is called convex, if for $x, y \in X$ and $t \in [0,1]$ holds that $f(tx + (1-t)y) \leq tf(x) + (1-t)f(y)$.*

- *The functional f is called strictly convex, if for $x, y \in X$, with $x \neq y$ and $t \in]0,1[$ holds that $f(tx + (1-t)y) < tf(x) + (1-t)f(y)$.*

Next a property, which is called lower semi-continuity, is introduced. This property is not restricted to convex functionals, but it might be used to investigate the differentiability properties of convex functionals and to prove the existence of minimizers. Further, a property is stated, which is equivalent to lower semi-continuity in case of convex functionals. This equivalence will play an important role in Chapter 3, where we investigate the discrepancy principle by Morozov.

Definition and Proposition 1.3.2 *Let X be a real Banach space and $f : X \to \mathbb{R} \cup \{-\infty, \infty\}$.*

1. *The functional f is called lower semi-continuous, if it satisfies the two equivalent conditions:*

   - *the sub-level sets $M_Q = \{x \in X \mid f(x) \leq Q\}$ are closed in X for all $Q \in \mathbb{R}$,*
   - *$f(\bar{x}) \leq \liminf_{x \to \bar{x}} f(x)$ for all $\bar{x} \in X$.*

2. *The functional f is called weakly lower semi-continuous, if it satisfies the two equivalent conditions:*

- the sub-level sets $M_Q = \{x \in X \mid f(x) \leq Q\}$ are weakly closed in X for all $Q \in \mathbb{R}$,

- $f(\bar{x}) \leq \liminf_{x \to \bar{x}} f(x)$ for all $\bar{x} \in X$.

A proof of the first equivalence result can be found in [49]. The second equivalence result can be proved in an analogous way.

Remark 1.3.3 *In the literature the above properties are sometimes called (weak) lower semi-continuity and (weak) sequential lower semi-continuity, respectively. In Banach spaces the indicated equivalences hold. Since we will always consider Banach or Hilbert spaces it will be referred to these equivalent properties in the following as (weak) lower semi-continuity.*

Moreover, in case of convex functionals both lower semi-continuity and weak lower semi-continuity are equivalent.

Proposition 1.3.4 *Let X be a real Banach space and $f : X \to \mathbb{R} \cup \{-\infty, \infty\}$ be a convex functional. Then the functional f is lower semi-continuous if and only if it is weakly lower semi-continuous.*

Proof. First we show that every convex closed set is also weakly closed.

Let M be a convex and closed set in X. We pick a sequence $\{x_n\}_n$ in M, with $x_n \rightharpoonup x$. Assume $x \notin M$, then by the Hahn-Banach separation theorem $x^* \in X^*$ and $\alpha \in \mathbb{R}$ exist such that

$$\langle y, x^* \rangle \; \leq \; \alpha \quad \text{for all } y \in M$$
$$\text{and} \quad \langle x, x^* \rangle \; > \; \alpha.$$

Since $\{x_n\}_n$ converges weakly, this implies $\langle x, x^* \rangle \leq \alpha$ and forms a contradiction.

Since f is assumed to be convex, the sub-level sets are convex as well. Further, if all sub-level sets are assumed to be closed, they are also weakly closed. This proves the first direction of the statement. The second implication is clear, since every convergent sequence is also weakly convergent and therefore weak lower semi-continuity implies lower semi-continuity. ∎

To be able to minimize functionals, we have to deal with the derivative of functionals. In the following definition some concepts are summarized, which hold independent of the convexity of the considered functional.

Definition 1.3.5 *Let X be a real Banach space and $f : X \to \mathbb{R} \cup \{-\infty, \infty\}$.*

- *The directional derivative of f at x in direction h is defined by*

$$f'(x; h) = \lim_{t \searrow 0} \frac{f(x+th) - f(x)}{t},$$

if the limit exists.

- If a $x^* \in X^*$ exists such that

$$f'(x; h) = \langle x^*, h \rangle \qquad \text{for all } h \in X,$$

  then f is called Gâteaux differentiable at x and $f'(x) = x^*$ is called the Gâteaux derivative of f at x.

- If a $x^* \in X^*$ exists such that an expansion of the form

$$f(x + h) = f(x) + \langle x^*, h \rangle + o(\|h\|) \qquad \text{as } \|h\| \to 0$$

  holds for all h in a neighborhood of zero, then f is called Fréchet differentiable at x and $f'(x) = x^*$ is called the Fréchet derivative of f at x.

For convex functionals, which are not differentiable in the sense of Gâteaux or Fréchet, a more general concept can be defined, the so-called subdifferential. Although the subdifferential can also be defined for non-convex functionals, only a definition for the convex case is presented, since we will just need this definition. The subdifferential of a functional may be a set-valued mapping as we will see in the following definition.

Definition 1.3.6 Let X be a real Banach space and $f : X \to \mathbb{R} \cup \{-\infty, \infty\}$ be a convex functional. An element $x^* \in X^*$ is called a subgradient of f in x, if and only if $f(x) \neq \pm\infty$ and

$$f(y) \geq f(x) + \langle x^*, y - x \rangle \qquad \text{for all } y \in X.$$

The set of all subgradients of f in x is called subdifferential and denoted by $\partial f(x)$.

Example 1.3.7 As a simple example we have a look at $f : \mathbb{R} \to \mathbb{R}$ with $f(x) = |x|$. In this case the subdifferential is given by

$$\partial f(x) = \begin{cases} -1 & : \ x < 0 \\ [-1, 1] & : \ x = 0 \\ 1 & : \ x > 0 \end{cases}.$$

See Figure 1.1 for an illustration.

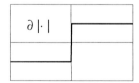

Figure 1.1: The absolute value and its subgradient.

As mentioned above, the subdifferential of a convex function at a specific point may contain more than one element. Therefore the concept of set-valued functions is introduced, see, e.g., [49]. We can characterize a mapping f which maps elements of a set X to subsets of a second set Y by its graph:

$$f(x) = \{y \in Y \mid (x, y) \in \text{graph}(f)\},$$

where the graph is defined by

$$\text{graph}(f) = \{(x, y) \in X \times Y \mid y \in f(x)\}.$$

This characterization in turn gives us the possibility to define the inverse of a set-valued function in a simple way. The inverse is defined by

$$f^{-1}(y) = \{x \in X \mid (x, y) \in \text{graph}(f)\}.$$

In the following, set-valued functions f will be indicated by $f : X \rightrightarrows Y$.

Now, we turn back to subdifferentials. To calculate subdifferentials of Tikhonov functionals, we need some subdifferential calculus. In the following, important rules for the computation of subdifferentials are summarized which can be found, e. g. in [54].

Proposition 1.3.8 *Let X be a real Banach space, $f : X \to \mathbb{R} \cup \{-\infty, \infty\}$ and $\lambda > 0$, then for all $x \in X$ it holds*

$$\partial(\lambda f)(x) = \lambda \partial f(x).$$

Proposition 1.3.9 (Moreau, Rockafellar) *Let X be a real Banach space and let $f, g : X \to \mathbb{R} \cup \{\infty\}$ be convex. If there is a point $\bar{x} \in D(f) \cap D(g)$, where f is continuous, then for all $x \in X$ it holds*

$$\partial(f + g)(x) = \partial f(x) + \partial g(x).$$

Proposition 1.3.10 *Let X and Y be real Banach spaces, $A : X \to Y$ be a bounded linear operator and $f : Y \to \mathbb{R} \cup \{\infty\}$ be convex and lower semi-continuous. If there is a point $A\bar{x}$, where f is continuous and finite, then for all $x \in X$ it holds*

$$\partial(f \circ A)(x) = A^* \partial f(Ax).$$

To connect differential calculus and subdifferential calculus, a statement is presented which connects the Gâteaux derivative and the subdifferential to each other.

Proposition 1.3.11 *Let X be a real Banach space and $f : X \to \mathbb{R} \cup \{-\infty, \infty\}$ be convex. If f is Gâteaux differentiable at $x \in X$, it is subdifferentiable at x and $\partial f(x) = \{f'(x)\}$. Conversely, if f is continuous and finite at $x \in X$ and has a unique subgradient, then f is Gâteaux differentiable at x and $f'(x)$ is the unique subgradient in $\partial f(x)$.*

As mentioned above, in the upcoming parts of this thesis we will aim at solving certain minimization problems with respect to convex functionals. Hence, the next proposition provides us with a basic statement for the investigation of convex minimization problems in real Banach spaces. This statement is essential for the investigation of the minimization of Tikhonov-type functionals. See [54] for a proof.

Proposition 1.3.12 *Let $f : X \to \mathbb{R} \cup \{\infty\}$ be a convex and proper functional, i. e.*
$f \not\equiv \infty$, *on the real Banach space X, then x is a solution of*

$$\inf_{x \in X} f(x) = \alpha$$

if and only if

$$0 \in \partial f(x).$$

Besides the outlined results from convex analysis, we also need some concepts from Banach space theory, especially results describing the geometry of those spaces. This is the topic of the next section.

1.4   Geometry of Banach spaces

For our investigation of Morozov's discrepancy principle in Chapter 3, we will need some results on the geometrical properties of Banach spaces. We will need uniform convexity of Banach spaces and convexity of power type, as well as the concept of duality mappings and Bregman distances. In the current section a brief introduction only to those concepts, which are needed in this thesis, is given. For a detailed introduction to this subject the interested reader is referred, e. g. to [13, 31].

To start with the degree of convexity of a Banach space X is described. To measure the degree of convexity, the modulus of convexity is introduced.

Definition 1.4.1 *Let X be a Banach space. The function $\delta_X : [0, 2] \to [0, 1]$ defined by*

$$\delta_X(\varepsilon) = \inf\{1 - \|\tfrac{1}{2}(x + y)\| \mid \|x\| = \|y\| = 1, \|x - y\| \geq \varepsilon\}$$

is called modulus of convexity of X.

Figure 1.2 illustrates this definition for the two dimensional case. Based on this modulus of convexity, an important property of a Banach space can be defined, namely the so-called uniform convexity.

Definition 1.4.2 *A Banach space X is called uniformly convex, if*

$$\delta_X(\varepsilon) > 0 \quad \text{for any } \varepsilon \in]0, 2].$$

This means especially that the curvature of the unit ball in X is strictly positive.

The previous definition provides us just with a qualitative convexity criterion. To distinguish several degrees of convexity, the so-called q-convexity or convexity of power type can be defined.

Definition 1.4.3 *A Banach space X is called q-convex, if there is a constant $c > 0$ such that*

$$\delta_X(\varepsilon) \geq c\varepsilon^q \quad \text{for any } \varepsilon \in]0, 2].$$

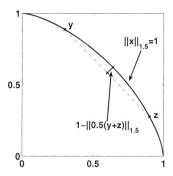

Figure 1.2: The modulus of convexity for $X = \ell_{1.5}$.

There are many well-known spaces which fit into this framework as we can see in the following example.

Example 1.4.4 *Note that for the sequence spaces ℓ_r as well as for the Lebesgue spaces $L_r(\Omega)$ and Sobolev spaces $W^{m,r}(\Omega)$, in each case with an open, bounded Lipschitz domain $\Omega \in \mathbb{R}^n$, holds*

$$\ell_r, L_r(\Omega), W^{m,r}(\Omega) \text{ with } \begin{cases} 1 < r \le 2 & \text{are} \quad \text{2-convex} \\ 2 \le r < \infty & \text{are} \quad \text{r-convex,} \end{cases}$$

see [26, 52].

At this point it has to be remarked that Ω is used for different mathematical objects. However, the meaning should arise out of the context in each case.

Another important concept in the theory of Banach spaces is the idea of duality mappings. Duality mappings connect a Banach space with its dual space. The mappings may be set-valued and can be understood as subdifferentials as we will see in the following.

Definition 1.4.5 *Let X be a Banach space and $1 < q < \infty$, then the set-valued mapping $J_q : X \rightrightarrows X^*$ defined by*

$$J_q(x) = \{x^* \in X^* \mid \langle x^*, x \rangle = \|x\|\|x^*\| \,, \, \|x^*\| = \|x\|^{q-1}\}$$

is called the duality mapping of X with weight function $t \mapsto t^{q-1}$. A single-valued selection of J_q is denoted by j_q.

The following theorem connects duality mappings to subdifferentials and can be found in [13, Theorem 4.4].

Theorem 1.4.6 (Asplund) *Let $1 < q < \infty$, then for the norm of a Banach space X it holds*

$$J_q(x) = \partial\{\tfrac{1}{q}\|x\|_X^q\} \quad \text{for each } x \in X.$$

Now, we are equipped with all necessary definitions to state a fundamental theorem proved by Xu and Roach. For uniformly convex Banach spaces it provides us with a characteristic inequality, which is essential to connect convergence in norm and in Bregman distance with each other as we will see in Chapter 3.

Theorem 1.4.7 *[52, Theorem 1] Let $1 < q < \infty$, then the following statements are equivalent:*

1. *X is a uniformly convex Banach space.*

2. *For any $x, y \in X$ it holds*

$$\|x - y\|_X^q \geq \|x\|_X^q - q \langle j_q(x), y \rangle + \sigma_q(x, y), \tag{1.7}$$

with

$$\sigma_q(x, y) = q K_q \int_0^1 \frac{(\max\{\|x - ty\|_X, \|x\|_X\})^q}{t} \delta_X \left(\frac{t\|y\|_X}{2\max\{\|x - ty\|_X, \|x\|_X\}} \right) dt,$$

where j_q denotes an arbitrary single-valued selection for the duality mapping J_q and K_q is a positive constant only depending on q.

Remark 1.4.8 *The statement of the previous theorem is a generalization of the well-known polarization identity*

$$\|x - y\|^2 = \|x\|^2 - 2 \langle x, y \rangle + \|y\|^2 \tag{1.8}$$

in real Hilbert spaces. To clarify this claim it would be necessary to introduce further geometric properties of Banach spaces, the so-called uniform smoothness and p-smoothness, see [52] for details. These properties lead to a second inequality

$$\|x - y\|_X^p \leq \|x\|_X^p - p \langle j_p(x), y \rangle + \tilde{\sigma}_p(x, y). \tag{1.9}$$

Moreover, investigations of equality (1.8) and inequalities (1.7) and (1.9) justify that every real Hilbert space is 2-convex and 2-smooth. These smoothness properties are mentioned for completeness, but detailed definitions are omitted since we do not need them in this thesis.

Next the concept of Bregman distances is presented. Whereas in Hilbert spaces, convergence properties are in general stated in terms of the norm of the underlying space, it turns out that in case of Banach spaces it may often be more appropriate to use so-called Bregman distances for convergence analysis. This idea goes back to Bregman [10].

The definition of Bregman distances can be done in slightly different ways, see, e. g., [11, 12, 47]. We use the set-valued version also used by Burger and Osher [11], since this is the most qualified one for our purposes.

Definition 1.4.9 *Let X be a Banach space and $f : X \to \mathbb{R} \cup \{\infty\}$ be proper and convex. The set*

$$\Delta_f(x, y) = \{f(x) - f(y) - \langle y^*, x - y \rangle \mid y^* \in \partial f(y)\}$$

defines the Bregman distance of x and y for all $x, y \in D(f)$, for which $\partial f(y) \neq \emptyset$.

Note that the convex functional f may be the power of a norm, especially in case of a Hilbert space with $f(x) = \frac{1}{2}\|x\|^2$ it holds $\Delta_f(x,y) = \{\frac{1}{2}\|x - y\|^2\}$.

As we can see from the definition, the Bregman distance measures the distance between the convex functional f evaluated in x and its linearization at a point y, evaluated in x as well. Since the functional f is convex, the Bregman distance is always non-negative. Figure 1.3 illustrates this observation.

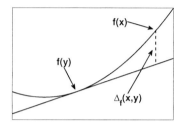

Figure 1.3: The Bregman distance.

Bregman distances are an essential tool when convergence rates of Tikhonov-type regularization schemes, including convex penalty terms, are investigated.

Finally, a property is presented, which is related to the Bregman distance. For special functionals the Bregman distance can be estimated from below in terms of the norm of the underlying Banach space. In that case, the functionals are called q-convex.

Definition 1.4.10 *Let $\Omega : X \to \mathbb{R} \cup \{\infty\}$ be proper and convex and X be a Banach space. Then Ω is called q-convex in v with $2 \leq q < \infty$, if for all $M > 0$ and $v^* \in \partial\Omega(v)$ there is a constant $0 < c < \infty$ such that for all $u \in X$, with $\|u - v\|_X \leq M$ it holds*

$$c\|u - v\|_X^q \leq \Omega(u) - \Omega(v) - \langle v^*, u - v \rangle. \tag{1.10}$$

We learned that as far as Banach spaces are concerned, Bregman distances can be used to prove convergence results. These distances can be seen as a generalization of the squared norm distances, since Bregman distances coincide with squared norm distances in case of Hilbert spaces. The property defined in Definition 1.4.10 shows that Bregman distances may for special Banach spaces be estimated from below by a power of the norm distance. This property will be essential to improve convergence rates in the chapter on Morozov's discrepancy principle.

The next section deals with a special Tikhonov-type regularization method called regularization with sparsity constraints. Some of the concepts introduced in the previous sections will play an important role while investigating this method.

1.5 Sparsity and iterated soft-shrinkage

In the following, we will discuss a regularization scheme, which has been studied intensively in recent years within the inverse problems community, namely regularization with sparsity constraints. It turned out that solutions of many inverse problems, for example inverse problems in signal or image processing or parameter identification problems, have a sparse structure. This means, the solution $x^\dagger$ of an inverse problem represented by an operator equation

$$Ax = y,$$

with $A : H \to \bar{H}$ mapping between real separable Hilbert spaces, can be expanded as

$$x^\dagger = \sum_{\lambda \in \Lambda^\dagger} \left\langle x^\dagger, \varphi_\lambda \right\rangle \varphi_\lambda,$$

where $\{\varphi_\lambda\}_{\lambda \in \Lambda}$ denotes some basis or frame of H, with index set Λ. The subset $\Lambda^\dagger \subset \Lambda$ denotes a finite index set containing those indices which correspond to the support of $x^\dagger$.

The pioneering paper [16] exploiting this sparsity assumption has been published by Daubechies, Defrise and De Mol in 2004. The authors formulated a regularization method based on Tikhonov regularization, where the classical quadratic penalty term was substituted by some sparsity enforcing penalty term of the form

$$\Omega(x) = \|x\|_{\mathbf{w},p}^p = \sum_{\lambda \in \Lambda} w_\lambda |x_\lambda|^p = \sum_{\lambda \in \Lambda} w_\lambda |\langle x, \varphi_\lambda \rangle|^p, \qquad (1.11)$$

with some positive weights $w_\lambda \geq \omega > 0$, $1 \leq p \leq 2$ and an orthonormal basis $\{\varphi_\lambda\}_\lambda$ of H. For the rest of this section we adopt this definition as a general assumption.

Sparsity enforcing can be understood in the following sense: assume a Tikhonov-type functional

$$\Gamma_{\alpha\mathbf{w},p,\delta}(x) = \|Ax - y^\delta\|^2 + \alpha \|x\|_{\mathbf{w},p}^p. \qquad (1.12)$$

Investigating the penalty term, defined in (1.11), we see that it consists of a linear combination of elements $|x_\lambda|^p$ including the expansion coefficients x_λ of x with respect to the basis $\{\varphi_\lambda\}_\lambda$. Considering small coefficients with $|x_\lambda| < 1$, those coefficients are less damped for $p < 2$ than for $p = 2$. This in turn means that small coefficients are the more penalized the smaller p is chosen. Regarding the soft-shrinkage algorithm for minimizing (1.12), presented in the next paragraph, we observe that in case of $p = 1$ small coefficients are set to zero during the minimization process and therefore the minimization process leads to a sparse solution. In case of $p > 1$ small coefficients are shrinked and do not vanish in general.

Remark 1.5.1

- *Firstly, we can see Ω as a functional mapping the Hilbert space H to $\mathbb{R} \cup \{\infty\}$, but if we have a closer look at Ω it seems to be natural to define subspaces $B_{\mathbf{w},p} = \{z \in H \mid \|z\|_{\mathbf{w},p} < \infty\}$ of H. Then $\|\cdot\|_{\mathbf{w},p}$ as in (1.11) defines a norm on these subspaces. The spaces are Banach spaces and in case of $w_\lambda \to \infty$ they are compactly embedded in H, see, e. g., [9, 16]. For a bounded sequence of weights, the spaces are still continuously embedded in H.*

- For a special choice of the system $\{\varphi_\lambda\}_\lambda$, namely choosing a wavelet basis, and choosing the weights as $w_\lambda = 2^{\sigma p |\lambda|}$, where $\sigma = s + d(1/2 + 1/p) > 0$, the norm $\|\cdot\|_{\mathbf{w},p}$ forms an equivalent norm of the Besov space $B_{p,p}^s(\mathbb{R}^d)$. According to the notation in [16], the basis $\{\varphi_\lambda\}_\lambda$ contains wavelets up to a coarsest scale $j = 0$ as well as the scaling functions on that scale. Here and in the following we denote the scale of the wavelets by $|\lambda| = j$. See also [14, 18, 35] for details. The norms of Besov spaces are often defined via so-called moduli of smoothness. A proper definition of those spaces and norms needs some fundamental results from approximation theory, which go far beyond the scope of this thesis. The interested reader is referred, e. g. to [14, 18] for a detailed discussion of this topic.

Finally, we prove a fundamental estimate, which we will need several times in the following chapters and remark on some further properties of the functional Ω.

Lemma 1.5.2 For $\|\cdot\|_{\mathbf{w},p}$ as defined in (1.11) the following inequality holds

$$\|x\|_H \leq \frac{1}{\omega^{1/p}} \|x\|_{\mathbf{w},p}$$

for all $x \in H$.

Proof. Since $1 \leq p \leq 2$, we can estimate

$$1 = \sum_\lambda \left(\frac{|x_\lambda|}{\|x\|_H} \right)^2 \leq \sum_\lambda \left(\frac{|x_\lambda|}{\|x\|_H} \right)^p$$

and obtain

$$\|x\|_H^p \leq \sum_\lambda |x_\lambda|^p \leq \frac{1}{\omega} \sum_\lambda w_\lambda |x_\lambda|^p = \frac{1}{\omega} \|x\|_{\mathbf{w},p}^p, \qquad (1.13)$$

which proves the statement. ■

In the following, the main aspects of regularization with sparsity constraints are summarized, following the lines of [16]. As mentioned above, we consider Tikhonov functionals of the form (1.12). Hence, we have to treat two main tasks. First we will introduce an algorithm for calculating minimizers of (1.12) and second we will justify that the proposed scheme, combining the minimization of Tikhonov-type functionals and an appropriate parameter choice strategy, forms a regularization method. The first question is the topic of the next paragraph. The latter is treated in the paragraph after next.

1.5.1 Contraction property and minimization scheme

In this paragraph we take a look at an iterative minimization scheme for Tikhonov-type functionals $\Gamma_{\alpha\mathbf{w},p,\delta}$ defined in (1.12). Minimizing $\Gamma_{\alpha\mathbf{w},p,\delta}$ directly would mean to solve a coupled system of non-linear equations for the expansion coefficients x_λ. To overcome these problems, so-called surrogate functionals are introduced, for which the minimization is much easier. These functionals are defined by

$$\Phi_{\alpha\mathbf{w},p,\delta}(x;u) = \Gamma_{\alpha\mathbf{w},p,\delta}(x) - \|Ax - Au\|^2 + \|x - u\|^2, \qquad (1.14)$$

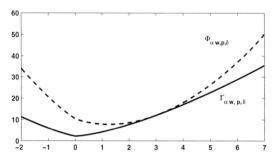

Figure 1.4: One dimensional example with surrogate functional $\Phi_{2.8,1.3,\delta}(x;3) = |0.3x - 2|^2 + 2.8|x|^{1.3} - |0.3x - 0.9|^2 + |x - 3|^2$.

where u denotes an arbitrary element of H. See Figure 1.4 for an illustration. Assuming $\|A\| < 1$ implies that the functionals are strictly convex and hence the minimizers are unique.

Remark 1.5.3 *The authors of [8] proved that the assumption $\|A\| < 2^{1/2}$ is sufficient to prove linear convergence of the minimization scheme presented below. Nevertheless, we assume $\|A\| < 1$ in the following, which is not restrictive, since the problem can always be rescaled to satisfy this assumption. To avoid rescaling of the problem, a constant C can be introduced in such a way that $\Gamma_{\alpha w,p,\delta}(x) - \|Ax - Au\|^2 + C\|x - u\|^2$ is convex with respect to x. The theory presented hereafter can then be done in an analogous way incorporating this constant.*

By rearrangement of the surrogate functionals we obtain

$$\Phi_{\alpha w,p,\delta}(x;u) = \sum_\lambda \left[x_\lambda^2 - 2x_\lambda(u - A^*(Au - y^\delta))_\lambda + \alpha w_\lambda |x_\lambda|^p \right]$$
$$+ \|y^\delta\|^2 + \|u\|^2 - \|Au\|^2.$$

The main advantage of this construction is that the coupling term $\|Ax\|^2$ cancels out. Obviously, minimizing $\Phi_{\alpha w,p,\delta}(\,\cdot\,;u)$ with a fixed u is equivalent to solving

$$\min_{x \in H} \sum_\lambda x_\lambda^2 - 2x_\lambda(u - A^*(Au - y^\delta))_\lambda + \alpha w_\lambda |x_\lambda|^p. \tag{1.15}$$

Since the coupling term has been removed, the minimizer of (1.15) can be calculated componentwise. For $p > 1$ every summand of (1.15) can be differentiated by x_λ and set to zero, which leads to the first order necessary conditions for a minimizer of (1.15)

$$x_\lambda + \frac{\alpha p w_\lambda}{2} \operatorname{sgn}(x_\lambda)|x_\lambda|^{p-1} = (u - A^*(Au - y^\delta))_\lambda \quad \text{for all } \lambda. \tag{1.16}$$

In case of $p = 1$ the summands of (1.15) are not differentiable in a classical sense, but since they are convex, we can apply the subdifferential calculus introduced in Section 1.3. Note that sgn denotes the set-valued sign function

$$\operatorname{sgn}(x) = \begin{cases} -1 & : \ x < 0 \\ [-1,1] & : \ x = 0 \\ 1 & : \ x > 0 \end{cases}.$$

We apply Proposition 1.3.9 and the first order necessary conditions (1.16) read as

$$0 \in \partial \left(x_\lambda^2 - 2x_\lambda(u - A^*(Au - y^\delta))_\lambda + \alpha w_\lambda |x_\lambda| \right)$$
$$\Leftrightarrow \quad 0 \in 2x_\lambda - 2(u - A^*(Au - y^\delta))_\lambda + \alpha w_\lambda \partial |x_\lambda|$$
$$\Leftrightarrow \quad (u - A^*(Au - y^\delta))_\lambda \in \left(I + \tfrac{\alpha w_\lambda}{2} \partial |\cdot| \right)(x_\lambda)$$
$$\Leftrightarrow \quad x_\lambda = \left(I + \tfrac{\alpha w_\lambda}{2} \partial |\cdot| \right)^{-1} ((u - A^*(Au - y^\delta))_\lambda)$$
$$\Leftrightarrow \quad x_\lambda = \mathrm{sgn}((u - A^*(Au - y^\delta))_\lambda)[|(u - A^*(Au - y^\delta))_\lambda| - \tfrac{\alpha w_\lambda}{2}]_+,$$

see also Proposition 1.3.12. Here $f_+(z)$ denotes the non-negative part of f, i. e.

$$f_+(z) = \left\{ \begin{array}{ccc} f(z) & : & f(z) \geq 0 \\ 0 & : & f(z) < 0 \end{array} \right. .$$

Note further that the mappings $\left(I + \tfrac{\alpha w_\lambda}{2} \partial |\cdot| \right)^{-1}$ are single-valued. See, e. g. [49] for a formal justification or consider Figure 1.5.

Finally, for $1 \leq p \leq 2$ the minimizer of (1.15) or (1.14) can be expressed by

$$\begin{aligned} x &= \sum_\lambda S_{\alpha w_\lambda, p}((u - A^*(Au - y^\delta))_\lambda) \varphi_\lambda \\ &= \mathbf{S}_{\alpha \mathbf{w}, p}(u - A^*(Au - y^\delta)) \end{aligned} \qquad (1.17)$$

containing the shrinkage functions

$$S_{\alpha w, p}(y) = \left\{ \begin{array}{lcc} \mathrm{sgn}(y)[|y| - \tfrac{\alpha w}{2}]_+ & : & p = 1 \\ G_{\alpha w, p}^{-1}(y) & : & 1 < p \leq 2 \end{array} \right. , \qquad (1.18)$$

where the

$$G_{\alpha w, p}(x) = x + \tfrac{\alpha w p}{2} \mathrm{sgn}(x) |x|^{p-1} \qquad (1.19)$$

define one-to-one mappings from $\mathbb{R}$ to itself. In the following, $\mathbf{S}_{\alpha \mathbf{w}, p}$ will be referred to as the soft-shrinkage operator. For an illustration of the shrinkage functions $S_{\alpha w, p}$ see Figure 1.5.

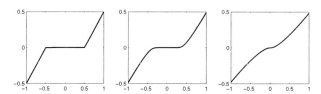

Figure 1.5: Shrinkage functions $S_{\alpha w, p}$ for $\alpha w = 1$ and $p = 1$, $p = 1.1$ and $p = 1.5$ (from left to right).

For $1 < p \leq 2$ the shrinkage functions and operators form contractions. This property is an essential tool for our investigation of adaptive soft-shrinkage in Chapter 2. Although this property was already mentioned in [16], it was not explicitly proved therein. Hence, we investigate the contraction property in more detail when proving the following lemma and corollary.

Lemma 1.5.4 *Let $1 < p \leq 2$ and $R < \infty$. The shrinkage function $S_{\alpha w, p} : [-R, R] \to \mathbb{R}$, defined in (1.18, 1.19) is a contraction with contraction constant*

$$q = q(\alpha, w, p, R) = \tfrac{1}{1+C} \quad \text{and} \quad C = C(\alpha, w, p, R) = \tfrac{\alpha w p (p-1)}{2} R^{p-2}.$$

In particular we obtain $|S_{\alpha w, p}(x)| \leq \tfrac{1}{1+C}|x|$ for all $x \in [-R, R]$.

Proof. Let $u, v \in [-R, R]$. The shrinkage function decreases the norm of its argument, hence $x = S_{\alpha w, p}(u)$ and $z = S_{\alpha w, p}(v)$ satisfy $x, z \in [-R, R]$.

For $1 < p \leq 2$ let $\tilde{C}$ denote a lower bound on the derivative of the function $\mathrm{sgn}(x)|x|^{p-1}$ on $[-R, R]$, i. e. $\tilde{C} = (p-1)R^{p-2}$. For all $x, z \in [-R, R]$ we obtain

$$\big|\, \mathrm{sgn}(x)|x|^{p-1} - \mathrm{sgn}(z)|z|^{p-1} \,\big| \geq \tilde{C}|x - z|.$$

By (1.16, 1.18, 1.19) x and z satisfy

$$x + \tfrac{\alpha w p}{2} \mathrm{sgn}(x)|x|^{p-1} = u,$$
$$z + \tfrac{\alpha w p}{2} \mathrm{sgn}(z)|z|^{p-1} = v.$$

Taking the difference of these two equations and observing that $x - z$ has the same sign as $\mathrm{sgn}(x)|x|^{p-1} - \mathrm{sgn}(z)|z|^{p-1}$ yields

$$
\begin{aligned}
|u - v| &= |x - z| + \tfrac{\alpha w p}{2} \big| \, \mathrm{sgn}(x)|x|^{p-1} - \mathrm{sgn}(z)|z|^{p-1} \,\big| \geq (1 + C)|x - z| \\
&= (1 + C)\, |S_{\alpha w, p}(u) - S_{\alpha w, p}(v)|,
\end{aligned}
$$

with $C = \tfrac{\alpha w p}{2}\tilde{C}$. We conclude that $S_{\alpha w, p}$ is a contraction on $[-R, R]$ with contraction constant $\tfrac{1}{1+C}$. The final statement of this lemma, $|S_{\alpha w, p}(x)| \leq \tfrac{1}{1+C}|x|$ for all $x \in [-R, R]$, follows by exploiting the contraction property as

$$|S_{\alpha w, p}(x)| = |S_{\alpha w, p}(x) - S_{\alpha w, p}(0)| \leq \tfrac{1}{1+C}|x - 0|.$$

∎

As a consequence of Lemma 1.5.4, the function-valued shrinkage operator $\mathbf{S}_{\alpha \mathbf{w}, p}$ is also a contraction on every bounded subset $\Omega \subset H$.

Corollary 1.5.5 *Let $1 < p \leq 2$ and let $\Omega = B(0, R) \subset H$ denote the ball of radius R in H. Then*

$$
\begin{aligned}
\mathbf{S}_{\alpha \mathbf{w}, p} : \quad \Omega &\longrightarrow \quad \Omega \\
x &\longmapsto \quad \sum_\lambda S_{\alpha w_\lambda, p}(x_\lambda)\varphi_\lambda
\end{aligned}
$$

is a contraction. The contraction constant $q = q(\alpha, \mathbf{w}, p, R) = \tfrac{1}{1+C}$ is the same as for the shrinkage function, see Lemma 1.5.4.

In particular we obtain $\|\mathbf{S}_{\alpha \mathbf{w}, p}(x)\| \leq \tfrac{1}{1+C}\|x\|$ for all $x \in \Omega$.

Proof. Let $x = \mathbf{S}_{\alpha\mathbf{w},p}(u)$ and $z = \mathbf{S}_{\alpha\mathbf{w},p}(v)$. Using the identity $\|x\|^2 = \sum_\lambda |x_\lambda|^2$ and applying the previous lemma to each coefficient of $x - z$ yields

$$\|x - z\|^2 = \sum_\lambda |x_\lambda - z_\lambda|^2 \le \left(\tfrac{1}{1+C}\right)^2 \sum_\lambda |u_\lambda - v_\lambda|^2 = \left(\tfrac{1}{1+C}\right)^2 \|u - v\|^2.$$

The final statement of this lemma, $\|\mathbf{S}_{\alpha\mathbf{w},p}(x)\| \le \tfrac{1}{1+C}\|x\|$ for all $x \in \Omega$, follows by exploiting the contraction property as

$$\|\mathbf{S}_{\alpha\mathbf{w},p}(x)\| = \|\mathbf{S}_{\alpha\mathbf{w},p}(x) - \mathbf{S}_{\alpha\mathbf{w},p}(0)\| \le \tfrac{1}{1+C}\|x - 0\|.$$

∎

Next the iterative minimization scheme for the original functional (1.12) proposed in [16] is presented:

$$\begin{aligned} x^0 \quad & \text{arbitrary} \\ x^{n+1} = \quad & \mathbf{S}_{\alpha\mathbf{w},p}(x^n - A^*(Ax^n - y^\delta)). \end{aligned} \tag{1.20}$$

This procedure corresponds to an alternating minimization of $\Phi_{\alpha\mathbf{w},p,\delta}(\,\cdot\,;\,\cdot\,)$ with respect to the first and the second argument. To justify the convergence of the iterative scheme a convergence result is quoted.

Theorem 1.5.6 *[16, Theorem 3.1] Let $A : H \to \bar{H}$ be a bounded linear operator mapping between real separable Hilbert spaces, with $\|A\| < 1$. Take $p \in [1,2]$, and let $\mathbf{S}_{\alpha\mathbf{w},p}$ be the shrinkage operator defined in (1.17), where the sequence $\mathbf{w} = \{w_\lambda\}_\lambda$ is uniformly bounded below away from 0. Then the sequence of iterates*

$$x^{n+1} = \mathbf{S}_{\alpha\mathbf{w},p}(x^n - A^*(Ax^n - y^\delta)), \quad n = 1, 2, \ldots,$$

with x^0 arbitrarily chosen in H, converges strongly to a minimizer of the functional $\Gamma_{\alpha\mathbf{w},p,\delta}$ defined in (1.12). If either $p > 1$ or $N(A) = \{\mathbf{0}\}$, then the minimizer x_α^δ of $\Gamma_{\alpha\mathbf{w},p,\delta}$ is unique, and every sequence $\{x^n\}_n$ of iterates converges strongly to x_α^δ, regardless of the choice of x^0.

The proof contained in [16] is quite technical, since the authors concentrate on the complicated case $p = 1$. In case of $1 < p < 2$, the proof can be done exploiting the contraction property of the shrinkage operators and gets much simpler.

Besides this general convergence result, Bredies and Lorenz proved the linear convergence of this iteration scheme in [8]. In contrary to the convergence proof in [16], which is based essentially on the non-expansivity of the soft-shrinkage operator, their proof is based on the decrease of the functional values and requires the so-called finite basis injectivity property (FBI-property).

Definition 1.5.7 *Let the operator $A : H \to \bar{H}$ be mapping between real separable Hilbert spaces and let $\{\varphi_\lambda\}_\lambda$ be an orthonormal basis of H. The operator A has the finite basis injectivity property, if for every subspace $U_\Lambda = \mathrm{span}\{\varphi_\lambda \mid \lambda \in \Lambda\} \subset H$ with $\Lambda \subset \mathbb{N}$ and $|\Lambda| < \infty$ the operator $A|_{U_\Lambda}$ is injective.*

The authors first prove a convergence rate for the distances of the functional values $\Gamma_{\alpha w,p,\delta}(x^n) - \Gamma_{\alpha w,p,\delta}(x_\alpha^\delta)$ and estimate these distances from below by $\|x^n - x_\alpha^\delta\|^2$ exploiting the FBI-property.

Remark 1.5.8 *Note that in case of $p = 1$ the FBI-property ensures the uniqueness of the minimizer of the Tikhonov functional $\Gamma_{\alpha w,1,\delta}$. See [8] for details.*

In the following it will be referred to the FBI-property at some point. This property can be used especially to improve convergence rates for Tikhonov-type regularization schemes as we will see in the upcoming parts of this thesis.

As mentioned in the introduction, besides iterated soft-shrinkage, many other methods, especially for the case $p = 1$ have been developed in recent years. We omit a detailed discussion at this point and refer the interested reader to the introduction and the references specified therein.

1.5.2   Regularization property and convergence rates

Next we investigate the regularization property of the proposed Tikhonov-type scheme. Therefore, we have to make assumptions on the dependence of the regularization parameter α on the noise level δ. In [16] the authors proved the regularizing property using the following requirements of the regularization parameter $\alpha = \alpha(\delta)$:

$$\lim_{\delta \to 0} \alpha = 0 \quad \text{and} \quad \lim_{\delta \to 0} \frac{\delta^2}{\alpha} = 0. \tag{1.21}$$

This is a classical assumption in regularization theory, see, e. g. Theorem 1.2.5. Choosing the regularization parameter according to (1.21) can be seen as an a priori choice of α depending on δ. Considering an a posteriori parameter choice for Tikhonov-type methods is the topic of Chapter 3.

For the a priori case we state the following theorem.

Theorem 1.5.9 *[16, Theorem 4.1] Assume $A : H \to \bar{H}$ to be a bounded, linear operator between real separable Hilbert spaces with $\|A\| < 1$. Let $1 \leq p \leq 2$ and $w_\lambda \geq \omega > 0$ for all λ. Assume that either $p > 1$ or $N(A) = \{\mathbf{0}\}$. For any $y^\delta \in \bar{H}$ with $\|y - y^\delta\| \leq \delta$ and any $\alpha > 0$, define x_α^δ to be the minimizer of (1.12). If $\alpha = \alpha(\delta)$ is chosen such that the requirement (1.21) is satisfied, then it holds*

$$\lim_{\delta \to 0} \|x_\alpha^\delta - x^\dagger\| = 0,$$

where $x^\dagger$ is the unique minimum $\|\cdot\|_{w,p}$-norm solution of $Ax = y$.

This result shows only the regularizing property of the method, but nothing is said about convergence rates. To compare different regularization methods it is important to investigate how fast the regularized solutions x_α^δ tend to the true or generalized solution $x^\dagger$ with respect to decreasing noise levels.

In the following, some results on convergence rates with slightly different assumptions concerning the operator A, the weights w_λ and the solution $x^\dagger$ are summarized. All these results incorporate an a priori choice of the regularization parameter. As we will see in Chapter 2, the following convergence rate results can be transfered to the case, where the operator A is evaluated via an adaptive scheme. For the proofs of the upcoming theorems the reader is referrred to the given references, since they need some careful preparation and are quite technical at times.

Nevertheless, in Chapter 3 on Morozov's discrepancy principle, some of the following convergence rate results will be proved for that a posteriori parameter choice rule. In doing so, we use similar arguments as in the proofs for the a priori case.

The upcoming listing of convergence rate results is not complete, but the reader should get an impression of the recent research in this field. The survey starts with a result which was proved by Daubechies, Defrise and De Mol, see [16, Chapter 4.2]. This result exploits the fact that the penalty term (1.11) coincides with a Besov penalty term for a special choice of the basis $\{\varphi_\lambda\}_\lambda$ and weights w_λ, see also Remark 1.5.1. Further, the operator A has to satisfy a special smoothing property. This can be summarized in the following theorem.

Theorem 1.5.10 (Daubechies, Defrise, De Mol) *Let the listed assumptions be satisfied:*

- $A : L_2(\mathbb{R}^d) \to L_2(\mathbb{R}^d)$ *linear, bounded and invertible on its range, with* $\|A\| < 1$,

- $\{\varphi_\lambda\}_\lambda$ *orthonormal wavelet basis of* $L_2(\mathbb{R}^d)$,

- $w_\lambda = 2^{\sigma p |\lambda|}$, *with* $\sigma = s + d\,(1/2 - 1/p) > 0$, $1 \leq p \leq 2$ *and* $|\lambda|$ *denoting the scale of the wavelet, i. e.* $\|\cdot\|_{\mathbf{w},p}$ *coincides with the norm of the Besov space* $B^s_{p,p}(\mathbb{R}^d)$,

- $\|x^\dagger\|_{\mathbf{w},p} \leq \rho < \infty$,

- *a* $\mu > 0$ *and constants* $c = c(\sigma,\mu)$ *and* $C = C(\sigma,\mu)$ *exist such that*

$$c \sum_\lambda 2^{-2|\lambda|\mu}|h_\lambda|^2 \leq \|Ah\|^2 \leq C \sum_\lambda 2^{-2|\lambda|\mu}|h_\lambda|^2$$

for all $h \in L_2(\mathbb{R}^d)$,

- $\alpha = \frac{\delta^2}{\rho^p}$.

Then it holds

$$\|x^\delta_\alpha - x^\dagger\| = \mathcal{O}\left(\delta^{\frac{\sigma}{\sigma+\mu}}\right) \quad \text{for} \quad \delta \to 0.$$

A more general result, which does not only cover the sparsity case, was proved by Burger and Osher in 2004, see [11]. They proved convergence rates for Tikhonov regularization considering Tikhonov-type functionals

$$\Gamma_{\alpha,\delta}(x) = \|Ax - y^\delta\|^2 + \alpha\Omega(x). \tag{1.22}$$

The authors make the following assumptions. The operator $A : X \to Y$ is assumed to be a linear and continuous operator mapping from a Banach space X to a Hilbert space Y. Further, it is assumed that the operator A is also continuous with respect to a possibly weaker Topology $\mathcal{T}$ on X, i. e. $\mathcal{T}$ is contained in the topology induced by $\| \cdot \|_X$. Moreover, the functional $\Omega : X \to \mathbb{R} \cup \{\infty\}$ is supposed to be convex and lower semi-continuous with respect to the topology $\mathcal{T}$, with compact sub-level sets $M_Q = \{x \in X \mid \Omega(x) \le Q\}$ in $\mathcal{T}$, which are, in addition, non-empty for $Q \ge 0$.

This setting ensures basically the existence of minimizers for the considered Tikhonov-type functionals (1.22) and the regularizing property of these minimizers when the error level δ tends to zero.

The next remark explains how our sparsity approach fits into this framework.

Remark 1.5.11 *As we have seen in Remark 1.5.1, functionals $\Omega(\cdot) = \| \cdot \|_{\mathbf{w},p}$ are norms of subspaces $X = B_{\mathbf{w},p}$ of H. These Banach spaces are defined by $B_{\mathbf{w},p} = \{z \in H \mid \|z\|_{\mathbf{w},p} < \infty\}$. Further, we set $Y = \bar{H}$.*

Note that the norm topology on H induces a topology $\mathcal{T}$ on X which is weaker than the topology induced by $\| \cdot \|_{\mathbf{w},p}$. This can be seen, since $\|x\|_H \le \frac{1}{\omega^{1/p}} \|x\|_{\mathbf{w},p}$ holds true, see Lemma 1.5.2. This inequality implies also the continuity assumption on the operator $A : X \to \bar{H}$. Since $A : H \to \bar{H}$ is assumed to be continuous we obtain

$$\|Ax\|_{\bar{H}} \le \|A\|_{L(H,\bar{H})} \|x\|_H \le \|A\|_{L(H,\bar{H})} \frac{1}{\omega^{1/p}} \|x\|_X,$$

with $\|A\|_{L(X,\bar{H})} \le \|A\|_{L(H,\bar{H})} \frac{1}{\omega^{1/p}}$.

At least if the weights w_λ tend to infinity, i. e. for all $C > 0$ it holds that $|\{\lambda \in \Lambda \mid w_\lambda \le C\}| < \infty$, the embedding $\iota : X \to H$ is compact, see [16] or [9] for a proof. Especially, this compact embedding ensures that the sub-level sets M_Q are compact with respect to the weaker topology $\mathcal{T}$.

For the general framework, the convergence rate result [11, Theorem 2] reads as:

Theorem 1.5.12 *Let $\|y - y^\delta\| \le \delta$ and let $x^\dagger$ be an Ω-minimizing solution of $Ax = y$. In addition, assume that the source condition:*

a $\bar{z} \in Y$ exists such that $A^*\bar{z} \in \partial\Omega(x^\dagger)$

is satisfied. Further, if $\alpha \sim \delta$, then for each minimizer x_α^δ of the Tikhonov-type functional (1.22), a $d \in \Delta_\Omega(x_\alpha^\delta, x^\dagger)$ exists such that

$$d = \mathcal{O}(\delta) \quad for \quad \delta \to 0.$$

This theorem gives only a convergence rate in terms of Bregman distances $\Delta_\Omega(\cdot, \cdot)$ as defined in Definition 1.4.9. Considering operators mapping from Banach to Hilbert spaces, it seems to be impossible to end up with convergence results in the norm of the Banach space in general. To get convergence also in norm, the Banach space might satisfy some geometric properties, especially uniform convexity. In Chapter 3 we will investigate such situations in detail.

In the following, we specialize in the sparsity case. Based on the previous convergence rate result, the latest results published in 2008 by Lorenz [32] and Grasmair, Haltmeier and Scherzer [24] provide us with convergence rates for the sparsity case.

For the next theorems we always assume a bounded, linear operator A mapping between real separable Hilbert spaces with $\|A\| < 1$. Further, we assume for the noisy data $\|y - y^\delta\| \le \delta$, $w_\lambda \ge \omega > 0$ and denote by $x^\dagger$ the minimum $\| \cdot \|_{\mathbf{w},p}$-norm solution of $Ax = y$. For some of the results quoted below, it is necessary to presume additional assumptions on the operator and the solution of the operator equation. For instance, we will need the FBI-property of the operator, see Definition 1.5.7, and the sparsity of the solution $x^\dagger$, i. e. $x^\dagger = \sum_{\lambda \in \Lambda^\dagger} x_\lambda^\dagger \varphi_\lambda$ with $|\Lambda^\dagger| < \infty$.

Before the announced statements are presented, we prove a lemma, which focuses on an interesting connection between source conditions and sparsity of minimum norm solutions.

Lemma 1.5.13 *In case of $p = 1$ the source condition*

$$a \; \bar{z} \; \text{exists such that} \; A^* \bar{z} \in \sum_\lambda w_\lambda \operatorname{sgn}(x_\lambda^\dagger) \varphi_\lambda = \partial \|x^\dagger\|_{\mathbf{w},1}$$

implies that $x^\dagger$ is sparsely represented.

Proof. Set $x^* = A^* \bar{z}$ and define $\Lambda_\omega = \{\lambda \mid |x_\lambda^*| \geq \omega\}$. Since $x^* \in H$ we obtain

$$\infty > \|x^*\|_H^2 \geq \sum_{\lambda \in \Lambda_\omega} |x_\lambda^*|^2 \geq \omega^2 |\Lambda_\omega|,$$

which means that $|\Lambda_\omega| < \infty$. Further we have

$$x^* \in \partial \|x^\dagger\|_{\mathbf{w},1} \quad \Leftrightarrow \quad \sum_\lambda x_\lambda^* \varphi_\lambda \in \sum_\lambda w_\lambda \operatorname{sgn}(x_\lambda^\dagger) \varphi_\lambda,$$

which implies for all $\lambda \notin \Lambda_\omega$ that

$$|w_\lambda \operatorname{sgn}(x_\lambda^\dagger)| < \omega$$

and hence $x_\lambda^\dagger = 0$. Finally we conclude that $x^\dagger$ is sparse, since the index set Λ_ω is finite. ∎

Now a result by Lorenz [32] is presented.

Theorem 1.5.14 *Let $1 < p \leq 2$ and let $x^\dagger$ satisfy the source condition:*

$$a \; \bar{z} \in \bar{H} \; \text{exists such that} \; A^* \bar{z} = p \sum_\lambda w_\lambda \operatorname{sgn}(x_\lambda^\dagger) |x_\lambda^\dagger|^{p-1} \varphi_\lambda = \partial \|x^\dagger\|_{\mathbf{w},p}^p.$$

Then with the choice $\alpha \sim \delta$ it holds

$$
\begin{aligned}
\|A x_\alpha^\delta - y^\delta\| &= \mathcal{O}(\delta) &&\text{for} \;\; \delta \to 0, \\
\sum_\lambda w_\lambda |(x_\alpha^\delta - x^\dagger)_\lambda|^2 &= \mathcal{O}(\delta) &&\text{for} \;\; \delta \to 0, \\
\|x_\alpha^\delta - x^\dagger\| &= \mathcal{O}(\delta^{1/2}) &&\text{for} \;\; \delta \to 0.
\end{aligned}
$$

Assuming sparsity of the solution $x^\dagger$ and the FBI-property, the authors of [24] were able to improve the rates leading to the following theorem.

Theorem 1.5.15 *Let* $1 < p \leq 2$, $x^\dagger$ *be sparsely represented. Assume that the operator* A *obeys the FBI-property and let* $x^\dagger$ *satisfy the source condition:*

a $\bar{z} \in \bar{H}$ *exists such that* $A^*\bar{z} = p \sum_\lambda w_\lambda \operatorname{sgn}(x_\lambda^\dagger) |x_\lambda^\dagger|^{p-1} \varphi_\lambda = \partial \|x^\dagger\|_{w,p}^p.$

Then with the choice $\alpha \sim \delta$ *it holds*

$$\|x_\alpha^\delta - x^\dagger\| = \mathcal{O}(\delta^{1/p}) \quad \text{for} \quad \delta \to 0.$$

In case of $p = 1$ they were able to prove the following statement.

Theorem 1.5.16 *Let* $p = 1$ *and assume that the operator* A *obeys the FBI-property and let* $x^\dagger$ *satisfy the source condition:*

a $\bar{z} \in \bar{H}$ *exists such that* $A^*\bar{z} \in \sum_\lambda w_\lambda \operatorname{sgn}(x_\lambda^\dagger) \varphi_\lambda = \partial \|x^\dagger\|_{w,1}.$

Then with the choice $\alpha \sim \delta$ *it holds*

$$\|x_\alpha^\delta - x^\dagger\| = \mathcal{O}(\delta) \quad \text{for} \quad \delta \to 0.$$

At this point it has to be emphasized that the statements in [24] are formulated in a more general framework. Nevertheless, in our situation, the results reduce to the statements of Theorems 1.5.15 and 1.5.16.

Considering these results on convergence rates, it is striking that all results assume an a priori choice of the regularization parameter. However, in computational practice the regularization parameter is often chosen via an a posteriori strategy as we have seen in Chapter 1.2. We turn back to the question of a posteriori parameter choice for Tikhonov-type methods in Chapter 3.

The topic of the following chapter is an approach for the combination of adaptive operator evaluations and the minimization scheme (1.20). We assume that the operator A can not be evaluated exactly, but is evaluated via an adaptive scheme. We will see that minimizing (1.12) via iteration (1.20) also forms a regularization method in case of adaptive operator evaluations. Furthermore, we will justify that the presented results on convergence rates keep valid in the adaptive case.

Chapter 2

Iterated soft-shrinkage and inexact operator evaluations

Considering regularization methods for inverse problems formulated via linear operator equations

$$Ax = y \qquad (2.1)$$

it is usually assumed, at least for the theoretical analysis, that the bounded operator $A : H \rightarrow \bar{H}$ mapping between real separable Hilbert spaces with $\|A\| < 1$ can be evaluated exactly. In practice this is in general not the case. In fact there are always errors which occur due to discretization and numerical evaluation of the operator.

In the current chapter we investigate the case of regularization with sparsity constraints using the soft-shrinkage algorithm, introduced in Section 1.5.1, for the minimization of Tikhonov-type functionals

$$\Gamma_{\alpha\mathbf{w},p,\delta}(x) = \|Ax - y^\delta\|^2 + \alpha\|x\|_{\mathbf{w},p}^p \qquad (2.2)$$

with penalty terms defined in (1.11). Before we start our investigations, we take a look at two error models describing different discretization errors.

Any numerical method for calculating the solution of an operator equation (2.1) in infinite dimensional spaces requires a finite dimensional approximation of the operator A. Further, it may be useful to use a somewhat compressed version of A to simplify or accelerate computations. One canonical approach is to assume approximate operators A_h satisfying some error estimate of the form

$$\|A_h - A\| \leq h \qquad (2.3)$$

in operator norm. Especially, this means that the representation of A_h does not depend on the evaluation point x. For the classical Tikhonov regularization there are already some results taking this approach into account. We briefly discussed some of them in the introduction.

The former approach is designed for inverse problems based on operators which can be discretized in such a way that estimate (2.3) is computable. Thinking of adaptive schemes like finite element or wavelet Galerkin methods, which may also be used for evaluation, this estimate cannot be calculated. For example if the operator A is the

solution operator of a partial differential equation and the operator evaluations are real-
ized via an adaptive scheme, the representation of the operator may vary for every point
x. Therefore an estimate in operator norm cannot be available. A natural approach is
to assume a pointwise error estimate in this situation, i. e. given a precision h of the
adaptive scheme an estimate

$$\|[Ax]_h - Ax\| \leq h \tag{2.4}$$

holds for every x. A current result, which goes into the direction of adaptive operator
evaluations has been published by Ramlau, Teschke and Zhariy, see [42]. We discussed
it already in the introduction. Their approach combines a Landweber iteration and a
coarsening procedure to produce sparsely represented solutions.

Let us take a closer look at the two error concepts (2.3) and (2.4), because there are
some further differences which should be addressed at this point. As mentioned above,
in the adaptive case the realization of the adaptive evaluation scheme $[A \cdot]_h$ varies in
every point, which creates a non-linear scheme even in case of linear operators. In case
of operator approximations, linearity of the operators A_h is assumed.

Moreover, the different approaches require a different analysis of the regularization
methods. Analyzing regularization and convergence properties of a method considering
operator approximations A_h allows us to substitute the operator A directly by its approximate version. Thinking of Tikhonov regularization, it may be possible to do the
analysis without taking a special minimization scheme into account. Working with the
adaptive approach $[A \cdot]_h$ it is necessary to substitute exact operator evaluations by a
pointwise approximation. Hence, this requires to analyze the convergence properties in
connection with an explicitly given minimization routine for the Tikhonov functional.

In the upcoming part of this chapter we discuss the case of adaptive operator evaluations assuming an error estimate of the form (2.4) in combination with the soft-shrinkage
algorithm introduced in Section 1.5.1 for the minimization of Tikhonov-type functionals
defined in (2.2) with $1 < p \leq 2$. Besides the properties of the minimization scheme
itself, we also investigate the regularization property and convergence rates. At this
point it must be mentioned that essential parts of the first section of the current chapter
have already been published in [6]. However, the results presented in the following are
slightly improved and stated in a different way. The second part of this chapter briefly
handles the case of operator approximations using the error estimate (2.3), again only
for the case $1 < p \leq 2$. Finally, the last part deals with an approach for the iterated
soft-thresholding ($p = 1$). Unfortunately neither the regularizing property nor convergence rates are proved in this case. Hence, this part can be seen as an outlook to future
investigations.

2.1   Adaptive operator evaluations: $1 < p \leq 2$

In this section we start investigating the soft-shrinkage algorithm, see Section 1.5.1,

$$x^0 \quad \text{arbitrary}$$
$$x^{n+1} = \mathsf{S}_{\alpha w, p}(x^n - A^*(Ax^n - y^\delta)) \tag{2.5}$$

for the minimization of Tikhonov-type functionals of the form (2.2) assuming an adaptive
evaluation of the operators A and A^*. As mentioned above during this section we assume

that the adaptive evaluation schemes $[A \cdot]_h$ and $[A^* \cdot]_{h^*}$ fulfill pointwise error estimates of the form (2.4).

The key property while analyzing the adaptive soft-shrinkage method is the contraction property of the shrinkage operators $\mathbf{S}_{\alpha \mathbf{w}, p}$ with $1 < p \leq 2$. This property was already investigated in Section 1.5.1.

In case of $p = 1$ the shrinkage operator is just a non-expansive mapping, but not a contraction. In this case, the analysis requires different techniques and is handled separately in the last part of this chapter.

2.1.1 Adaptive soft-shrinkage iteration

This paragraph deals with the reformulation of the soft-shrinkage iteration (2.5) assuming adaptive operator evaluations. It should be emphasized again that in case of adaptive evaluations of the operators A and A^* the approximation of these operators is not fixed and might change in every point. In the following we assume adaptive schemes $[A \cdot]_h$ and $[A^* \cdot]_{h^*}$, which satisfy pointwise error estimates, i. e. for all x and z it holds

$$\|[Ax]_h - Ax\| \leq h \quad \text{and} \quad \|[A^*z]_{h^*} - A^*z\| \leq h^*.$$

To trace these adaptive operator evaluations back to the exact ones, we describe the adaptive operator evaluations as exact ones with an additional error term in the following form

$$
\begin{aligned}
[Ax]_h &= Ax + \eta^h \quad \text{with} \quad \|\eta^h\| \leq h \quad \text{and} \\
[A^*z]_{h^*} &= A^*z + \xi^{h^*} \quad \text{with} \quad \|\xi^{h^*}\| \leq h^*.
\end{aligned}
\tag{2.6}
$$

Note that the error terms η and ξ depend on the elements x and z respectively. In the following, it is avoided to indicate this fact, since we always deal with their norm estimates. Further, we assume $h = h^*$ for simplicity, otherwise the maximum of these to error bounds has to be used instead of h for the upcoming analysis.

Next we define the adaptive soft-shrinkage iteration by substituting the exact operator evaluations in iteration (2.5) by the adaptive ones. We obtain

$$
\begin{aligned}
x^0 &\quad \text{arbitrary} \\
x^{n+1} &= \mathbf{S}_{\alpha \mathbf{w}, p}(x^n - [A^*([Ax^n]_h - y^\delta)]_h).
\end{aligned}
\tag{2.7}
$$

Exploiting the equivalent formulation for the adaptive operator evaluations (2.6) we can reformulate iteration (2.7) using exact operator evaluations plus an additive error term.

Lemma 2.1.1 *We assume that an adaptive operator evaluation scheme satisfying (2.6) is given. Then iteration (2.7) can be reformulated as*

$$
\begin{aligned}
x^0 &\quad \text{arbitrary} \\
x^{n+1} &= \mathbf{S}_{\alpha \mathbf{w}, p}(x^n - A^*(Ax^n - y^\delta) - \chi^h),
\end{aligned}
\tag{2.8}
$$

where the error term χ^h satisfies $\|\chi^h\| \leq 2h$.

Proof. Substituting $[Ax]_h = Ax + \eta^h$ and $[A^*z]_h = A^*z + \xi^h$ in iteration (2.7), we obtain

$$
\begin{aligned}
x^{n+1} &= \mathbf{S}_{\alpha\mathbf{w},p}(x^n - [A^*([Ax^n]_h - y^\delta)]_h) \\
&= \mathbf{S}_{\alpha\mathbf{w},p}(x^n - [A^*(Ax^n + \eta^h - y^\delta)]_h) \\
&= \mathbf{S}_{\alpha\mathbf{w},p}(x^n - (A^*(Ax^n + \eta^h - y^\delta) + \xi^h)) \\
&= \mathbf{S}_{\alpha\mathbf{w},p}(x^n - A^*(Ax^n - y^\delta) - \chi^h)
\end{aligned}
$$

with $\chi^h = A^*\eta^h + \xi^h$ and $\|\chi^h\| \leq \|A^*\eta^h\| + \|\xi^h\| \leq 2h$.   ∎

Again we emphasize that an extra index n to indicate the dependence of χ^h on the iteration index will be omitted in the following, although this error term may vary in every iteration.

Remark 2.1.2 *Note that iteration (2.7) can also be interpreted as an exact soft-shrinkage iteration using perturbed iterates x_h^n and data y_h^δ, i. e.*

$$
\begin{aligned}
x^0 &\quad \text{arbitrary} \\
x^{n+1} &= \mathbf{S}_{\alpha\mathbf{w},p}(x_h^n - A^*(Ax_h^n - y_h^{\delta,n})),
\end{aligned}
$$

with $\|x_h^n - x^n\| \leq h$ and $\|y_h^{\delta,n} - y^\delta\| \leq 2h$. Further in case of summable error terms, i. e. $\sum_n \|\chi^{h,n}\| < \infty$, an iteration scheme like (2.8) was investigated in [15]. In our case this condition is not fulfilled in general.

2.1.2 Approaching the minimizer

In this paragraph, we assume that α and δ are fixed. We investigate the behavior of the iterates x^n produced by the adaptive iteration scheme (2.7). It will be shown that the iterates enter some ε-ball around the minimizer x_α^δ of (2.2), depending on a preassigned precision h, after a finite number of iteration steps. Note that strong convergence of the sequence of iterates can not be expected, since perturbations are added in every iteration step.

First we formulate a lemma, which describes the decrease of the distances between the iterates x^n and the minimizer x_α^δ.

Lemma 2.1.3 *Let $x^0 \in H$ be an arbitrary starting point. Let $\{x^n\}_{n\geq 0}$ denote the sequence of iterates produced by the soft-shrinkage iteration with adaptive operator evaluations (2.7). Define*

$$
\varepsilon^n = \|x^n - x_\alpha^\delta\| \quad \text{with } n \geq 0 \quad \text{and}
$$

$$
R^n = \varepsilon^{n-1} + \|x_\alpha^\delta\| + \|y^\delta\| + 2h \quad \text{with } n > 0.
$$

Denote the local contraction constant of $\mathbf{S}_{\alpha\mathbf{w},p}$ on $B(0, R^n)$ by

$$
\frac{1}{1+C^n} \tag{2.9}
$$

with

$$C^n = \frac{\alpha \omega p(p-1)}{2}(R^n)^{p-2} \tag{2.10}$$

depending on the iteration index n. With some given global precision h for the adaptive operator evaluations it holds then

$$x^n \in B(0, R^n) \quad \text{for } n > 0$$

and

$$\varepsilon^n \le \frac{2h}{1+C^n} + \frac{1}{1+C^n}\varepsilon^{n-1} \quad \text{for } n > 0.$$

Proof. We have to show that

$$x^n = \mathbf{S}_{\alpha w, p}(x^{n-1} - A^*(Ax^{n-1} - y^\delta) - \chi^h)$$

is located in $B(0, R^n)$ for every $n > 0$. Therefore we first show that $x^{n-1} - A^*(Ax^{n-1} - y^\delta) - \chi^h$ is located in $B(0, R^n)$ and then use the contraction property proved in Corollary 1.5.5 with the local contraction constant as defined in (2.9). Note first that

$$\|I - A^*A\| \le 1$$

since $\|A\| < 1$ and hence the following chain of inequalities holds

$$
\begin{aligned}
\|(I - A^*A)x\|^2 &= \|x\|^2 - 2\langle x, A^*Ax \rangle + \|A^*Ax\|^2 \\
&\le \|x\|^2 - 2\|Ax\|^2 + \|A\|^2\|Ax\|^2 \\
&< \|x\|^2 - \|Ax\|^2 \\
&\le \|x\|^2.
\end{aligned}
$$

Further we estimate

$$
\begin{aligned}
\|x^{n-1} - A^*(Ax^{n-1} - y^\delta) - \chi^h\| &\le \|(I - A^*A)(x^{n-1}) + A^*y^\delta - \chi^h\| \\
&\le \|x^{n-1}\| + \|y^\delta\| + 2h \\
&\le \|x^{n-1} - x_\alpha^\delta\| + \|x_\alpha^\delta\| + \|y^\delta\| + 2h \\
&= \varepsilon^{n-1} + \|x_\alpha^\delta\| + \|y^\delta\| + 2h \\
&= R^n,
\end{aligned}
$$

which implies by the contraction property $x^n \in B(0, R^n)$. The stated estimate for ε^n can now be proved with the help of the contraction property of $\mathbf{S}_{\alpha w, p}$. Since the minimizer x_α^δ happens to be a fixed point of iteration (2.5), see Theorem 1.5.6 and $x_\alpha^\delta - A^*(Ax_\alpha^\delta - y^\delta)$ is located in every ball $B(0, R^n)$, we obtain

$$
\begin{aligned}
\varepsilon^n &= \|x^n - x_\alpha^\delta\| \\
&= \|\mathbf{S}_{\alpha w, p}(x^{n-1} - A^*(Ax^{n-1} - y^\delta) - \chi^h) - \mathbf{S}_{\alpha w, p}(x_\alpha^\delta - A^*(Ax_\alpha^\delta - y^\delta))\| \\
&\le \frac{1}{1+C^n}\|(I - A^*A)(x^{n-1} - x_\alpha^\delta) - \chi^h\| \\
&\le \frac{1}{1+C^n}(\|x^{n-1} - x_\alpha^\delta\| + 2h) \\
&= \frac{2h}{1+C^n} + \frac{1}{1+C^n}\varepsilon^{n-1}. \tag{2.11}
\end{aligned}
$$

■

The previous lemma proves a decrease of the distances between the iterates x^n and the minimizer x_α^δ as long as h is small enough, i. e. $\frac{2h}{C^n} < \varepsilon^{n-1}$. Next we prove a useful estimate for these distances assuming a fixed precision h.

Lemma 2.1.4 *With the assumptions and notations of the previous lemma, a given precision h for the adaptive operator evaluations and as long as*

$$h < \frac{\varepsilon^{n-1}C^n}{2}$$

holds, we obtain a decreasing sequence $\{\varepsilon^n\}_n$ which satisfies the estimate

$$\varepsilon^n \leq \prod_{k=0}^{n-1} \left(\frac{2h}{\varepsilon^k} + 1\right) \left(\frac{1}{1+C^{k+1}}\right) \varepsilon^0.$$

Proof. With the help of the previous lemma we obtain

$$
\begin{aligned}
\varepsilon^n &\leq \frac{2h}{1+C^n} + \frac{1}{1+C^n}\varepsilon^{n-1} \\
&= \left(\frac{2h}{\varepsilon^{n-1}} + 1\right)\left(\frac{1}{1+C^n}\right)\varepsilon^{n-1} \\
&\leq \prod_{k=0}^{n-1} \left(\frac{2h}{\varepsilon^k} + 1\right)\left(\frac{1}{1+C^{k+1}}\right)\varepsilon^0.
\end{aligned}
$$

We see that the decrease of the distances ε^n can only be assured as long as

$$\left(\frac{2h}{\varepsilon^{n-1}} + 1\right)\left(\frac{1}{1+C^n}\right) < 1,$$

which is equivalent to

$$h < \frac{\varepsilon^{n-1}C^n}{2}. \qquad \blacksquare$$

The previous lemma gives ε^0 as an upper bound for the distances between the minimizer and the iterates.

Next we show that the iterates reach every ε-ball, which is larger than some $\varepsilon_{\min}$-ball, with $\varepsilon_{\min} = \varepsilon_{\min}(\alpha, \delta, h)$, after a finite number of iteration steps. To prove the next proposition we first state a technical lemma:

Lemma 2.1.5 *The functions*

$$F_p(x) = x^{\frac{p-1}{p-2}} + ax^{\frac{1}{p-2}} - b,$$

with $1 < p < 2$ and $a, b > 0$ are monotonically decreasing on $]0, \infty[$ and have a single root for some finite x.

Proof. The exponents $\frac{p-1}{p-2}$ and $\frac{1}{p-2}$ are negative for $1 < p < 2$, which means that $x^{\frac{p-1}{p-2}} + ax^{\frac{1}{p-2}}$ is monotonically decreasing towards zero. Subtracting a positive constant proves the statement. $\blacksquare$

Proposition 2.1.6 *Let the assumptions of Lemma 2.1.3 be valid and define two constants*

$$K_1 = K_1(\alpha, \delta) = \|x_\alpha^\delta\| + \|y^\delta\| \qquad \text{and} \qquad K_2 = K_2(\alpha, w, p) = \tfrac{\alpha w p(p-1)}{4}.$$

Let further ε_{min} be as follows:

- *for $1 < p < 2$ it is the unique solution of*

$$\varepsilon^{\frac{p-1}{p-2}} + (K_1 + 2h)\varepsilon^{\frac{1}{p-2}} - \left(\tfrac{h}{K_2}\right)^{\frac{1}{p-2}} = 0 \qquad (2.12)$$

- *and for $p = 2$ it is given by*

$$\varepsilon_{min} = \tfrac{2h}{\alpha w}. \qquad (2.13)$$

Then the iterates generated by the adaptive soft-shrinkage algorithm reach every ε-ball with $\varepsilon > \varepsilon_{min} > 0$ after a finite number of iteration steps.

Proof. As proved in Lemma 2.1.4, the distances ε^n form a decreasing sequence as long as the condition

$$h < \tfrac{\varepsilon^{n-1} C^n}{2} \qquad (2.14)$$

holds, with

$$C^n = 2K_2(R^n)^{p-2}$$

and

$$R^n = \varepsilon^{n-1} + K_1 + 2h.$$

Let us first treat the case of $1 < p < 2$. Inserting the definitions of C^n and R^n into condition (2.14) and substituting ε^{n-1} by ε leads to the following equivalent inequality

$$\varepsilon^{\frac{p-1}{p-2}} + (K_1 + 2h)\,\varepsilon^{\frac{1}{p-2}} - \left(\tfrac{h}{K_2}\right)^{\frac{1}{p-2}} < 0. \qquad (2.15)$$

Hence, we realize that we have a decrease of the ε^n as long as ε^{n-1} is larger than the unique solution ε_{min} of

$$\varepsilon^{\frac{p-1}{p-2}} + (K_1 + 2h)\varepsilon^{\frac{1}{p-2}} - \left(\tfrac{h}{K_2}\right)^{\frac{1}{p-2}} = 0.$$

In case of $p = 2$ condition (2.14) is equivalent to

$$h < \tfrac{\varepsilon^{n-1} \alpha w}{2}, \qquad (2.16)$$

since $C^n = \alpha w$ is independent of n. In that case, we obtain $\varepsilon_{min} = \tfrac{2h}{\alpha w}$.

The last step is to justify that the distances fall below every ε larger than ε_{min} after a finite number of iteration steps. The sequence $\{\varepsilon^n\}_n$ forms a monotonically decreasing sequence as long as ε^{n-1} satisfies (2.15) or (2.16), respectively or equivalently condition (2.14) is fulfilled.

Assume there is an index N such that (2.15) or (2.16) is violated for ε^N. That means $\varepsilon^N < \varepsilon_{\min}$.

On the other hand, assuming that (2.15) or (2.16) is satisfied for all n and $\varepsilon^n \to \tilde{\varepsilon}$, with $\tilde{\varepsilon} > \varepsilon_{\min} > 0$ means that there is no finite index N such that $\varepsilon^N < \tilde{\varepsilon}$.

In case of $p = 2$ the contraction constants $\frac{1}{1+C^n}$ are smaller than one and do not depend on n. This implies

$$\left(\tfrac{2h}{\varepsilon^n} + 1\right)\left(\tfrac{1}{1+C^{n+1}}\right) = \left(\tfrac{2h}{\varepsilon^n} + 1\right)\left(\tfrac{1}{1+\alpha\omega}\right) \to \left(\tfrac{2h}{\tilde{\varepsilon}} + 1\right)\left(\tfrac{1}{1+\alpha\omega}\right) < 1$$

and by Lemma 2.1.4 it follows $\varepsilon^n \to 0$ which forms a contradiction.

In case of $1 < p < 2$ the sequence $\{C^n\}_n$ is monotonically increasing with $C^n \to \tilde{C}$ and a finite $\tilde{C}$ depending on $\tilde{\varepsilon}$. Consequently it holds by (2.15) and (2.14)

$$\left(\tfrac{2h}{\varepsilon^n} + 1\right)\left(\tfrac{1}{1+C^{n+1}}\right) \to \left(\tfrac{2h}{\tilde{\varepsilon}} + 1\right)\left(\tfrac{1}{1+\tilde{C}}\right) < \tfrac{1+\tilde{C}}{1+\tilde{C}} = 1.$$

Again we get a contradiction, since by Lemma 2.1.4 it holds that $\varepsilon^n \to 0$.   ∎

The previous proposition guarantees that the iterates reach ε-balls around the minimizer x_α^δ after finitely many iteration steps depending on the parameters α, δ and the precision h. In addition to it, some further remarks should be made at this point.

Remark 2.1.7

1. *If there is an index N such that $\varepsilon^N < \varepsilon_{\min}$, the iterates will stay in this $\varepsilon_{\min}$-ball around x_α^δ since the following arguments hold. Assume an index $\tilde{N} > N$ with $\varepsilon^{\tilde{N}} \geq \varepsilon_{\min}$, this implies that (2.15) or (2.16) is satisfied for $\varepsilon^{\tilde{N}-1}$. By (2.11) we have then $\varepsilon_{\min} \leq \varepsilon^{\tilde{N}} \leq \varepsilon^{\tilde{N}-1}$. Applying the above argumentation recursively yields $\varepsilon_{\min} \leq \varepsilon^{\tilde{N}} \leq \varepsilon^{\tilde{N}-1} \leq \cdots \leq \varepsilon^N$, which forms a contradiction.*

2. *The previous proposition shows that there is a finite index $N = N(\alpha, \delta, h, \varepsilon)$ which ensures that $\varepsilon^N < \varepsilon$. Lemma 2.1.11 provides us with a concrete value for N.*

3. *In Lemma 2.1.8 it is shown that $\varepsilon_{\min} = \mathcal{O}(\frac{h}{\alpha})$, if h and α were chosen properly.*

4. *So far the estimates require the unknown minimizer x_α^δ. In practice this must be estimated by some known quantity. We comment on this fact in Remark 2.1.13.*

Next we investigate the regularizing property of the scheme.

2.1.3   Regularization property

So far we treated the minimization of Tikhonov functionals with fixed α and δ. To prove the regularization property, we have to investigate what happens when the noise level δ tends to zero. Especially, we have to investigate the dependence of $\varepsilon = \varepsilon_{\min}(\alpha, \delta, h)$ on the parameters α, δ and the precision h and how to couple these parameters to δ to get the convergence of ε to zero when δ goes to zero.

First we show that the radii ε tend to zero, when the noise level tends to zero and the parameters α and h are coupled to δ in an appropriate way.

Lemma 2.1.8 *Let the assumptions of Proposition 2.1.6 be valid and pick* $\alpha = \alpha(\delta)$ *such that*

$$\alpha \xrightarrow{\delta \to 0} 0 \quad \text{and} \quad \frac{\delta^2}{\alpha} \xrightarrow{\delta \to 0} 0$$

and further

$$h = \mathcal{O}(\alpha^\tau) \quad \text{for} \quad \alpha \to 0,$$

with $\tau > 1$. *Moreover, let* $\|y\| > 0$ *and* $\|y^\delta - y\| \leq \delta$. *Then it holds*

$$\varepsilon = \mathcal{O}(\tfrac{h}{\alpha}) \quad \text{which means} \quad \varepsilon \xrightarrow{\delta \to 0} 0.$$

Proof. Since for $p = 2$ the value ε is defined as

$$\varepsilon = \frac{2h}{\alpha w},$$

the statement is trivial.

In case $1 < p < 2$, we take a look at

$$\varepsilon^{\frac{p-1}{p-2}} + (K_1 + 2h)\varepsilon^{\frac{1}{p-2}} = (\tfrac{h}{K_2})^{\frac{1}{p-2}}.$$

We investigate the dependence of ε on α, δ and h. To this end, we reformulate the previous equality and obtain

$$(\varepsilon + \|x_\alpha^\delta\| + \|y^\delta\| + 2h)\varepsilon^{\frac{1}{p-2}} = C(\tfrac{h}{\alpha})^{\frac{1}{p-2}},$$

where C depends only on ω and p. Next we estimate the quantity $\varepsilon + \|x_\alpha^\delta\| + \|y^\delta\| + 2h$ from above. We start by investigating $\|x_\alpha^\delta\|$. With the help of Lemma 1.5.2 it follows:

$$
\begin{aligned}
\|x_\alpha^\delta\|^p &\leq \tfrac{1}{\omega}\|x_\alpha^\delta\|_{\mathbf{w},p}^p \\
&\leq \tfrac{1}{\alpha\omega}(\|Ax_\alpha^\delta - y^\delta\|^2 + \alpha\|x_\alpha^\delta\|_{\mathbf{w},p}^p) \\
&\leq \tfrac{1}{\alpha\omega}(\|Ax^\dagger - y^\delta\|^2 + \alpha\|x^\dagger\|_{\mathbf{w},p}^p) \\
&\leq \tfrac{1}{\omega}(\tfrac{\delta^2}{\alpha} + \|x^\dagger\|_{\mathbf{w},p}^p).
\end{aligned}
$$

Further we obtain

$$\|y^\delta\| \leq \|y - y^\delta\| + \|y\| \leq \delta + \|y\|.$$

We fix some δ^*, $\alpha^*(\delta^*)$ and $h^*(\alpha^*)$. By assumption $\frac{\delta^2}{\alpha} \to 0$ as $\delta \to 0$. This means for small δ, α and h it holds that

$$\left(\varepsilon + \left(\tfrac{1}{\omega}\left(\tfrac{(\delta^*)^2}{\alpha^*} + \|x^\dagger\|_{\mathbf{w},p}^p\right)\right)^{\frac{1}{p}} + \|y\| + \delta^* + 2h^*\right)\varepsilon^{\frac{1}{p-2}} \geq C\left(\tfrac{h}{\alpha}\right)^{\frac{1}{p-2}}$$

and since $\frac{h}{\alpha} \to 0$ as $\alpha \to 0$ the right hand side goes to infinity for $\alpha \to 0$. Together this implies that ε has to go to zero as well. We can fix an ε^* and substitute the ε in the bracket term by ε^* which finally leads to

$$\varepsilon \leq \bar{C}\tfrac{h}{\alpha}$$

with a finite constant $\bar{C} > 0$. ∎

Remark 2.1.9 *It can also be shown that there is a positive constant c such that $\varepsilon \geq c\frac{h}{\alpha}$.*

In general the $\varepsilon = \varepsilon_{\min}$-balls may not be reached after a finite number of iteration steps. Nevertheless, we know from Proposition 2.1.6 that every ε_θ-ball with $\varepsilon_\theta > \varepsilon$ will be reached after a finite number $N = N(\delta)$ of iteration steps.

Remark 2.1.10 *Note that the convergence properties do not change if we substitute ε by $\varepsilon_\theta = \theta\varepsilon$ with some fixed $\theta > 1$.*

Next, we give a concrete value for the maximal number of iteration steps needed to enter some given ε_θ-ball considering a fixed set $\{\delta, \alpha, h\}$.

Lemma 2.1.11 *With the assumptions of Proposition 2.1.6, a fixed collection of parameters $\{\delta, \alpha, h\}$ and*

$$\theta > \left(\frac{\varepsilon^0 + \|x_\alpha^\delta\| + \|y^\delta\| + 2h}{\varepsilon + \|x_\alpha^\delta\| + \|y^\delta\| + 2h} \right)^{2-p} \tag{2.17}$$

an ε_θ-ball around x_α^δ with $\varepsilon_\theta = \theta\varepsilon$ is reached after at most N iteration steps where N is given by

$$N = \left\lceil \frac{\log\left(\frac{\varepsilon_\theta}{\varepsilon^0}\right)}{\log(q)} \right\rceil$$

with

$$q = \left(\frac{2h}{\varepsilon_\theta} + 1 \right) \left(\frac{1}{1+C^1} \right).$$

Proof. Considering Lemma 2.1.4, since the C^n are monotonically increasing and as long as $\varepsilon^{n-1} \geq \varepsilon_\theta$ we see that

$$
\begin{aligned}
\varepsilon^n &\leq \prod_{k=0}^{n-1} \left(\frac{2h}{\varepsilon^k} + 1 \right) \left(\frac{1}{1+C^{k+1}} \right) \varepsilon^0 \\
&\leq \left[\left(\frac{2h}{\varepsilon_\theta} + 1 \right) \left(\frac{1}{1+C^1} \right) \right]^n \varepsilon^0 \\
&= q^n \varepsilon^0.
\end{aligned}
$$

In case of $1 < p < 2$ we use the definitions of C^1 and ε, see (2.10) and (2.12), respectively and obtain

$$\frac{C^1 \varepsilon_\theta}{2} = \left(\frac{\varepsilon^0 + \|x_\alpha^\delta\| + \|y^\delta\| + 2h}{\varepsilon + \|x_\alpha^\delta\| + \|y^\delta\| + 2h} \right)^{p-2} \theta h.$$

This yields

$$\left(\frac{\varepsilon^0 + \|x_\alpha^\delta\| + \|y^\delta\| + 2h}{\varepsilon + \|x_\alpha^\delta\| + \|y^\delta\| + 2h} \right)^{2-p} < \theta \implies h < \frac{C^1 \varepsilon_\theta}{2}$$
$$\implies q < 1.$$

If $p = 2$ condition (2.17) reduces to $1 < \theta$. Further, by (2.10) and (2.13) we obtain

$$h < \frac{C^1 \varepsilon_\theta}{2} = \theta h \implies q < 1.$$

Now, we are looking for an index n satisfying

$$\varepsilon_\theta \geq q^n \varepsilon^0,$$

which is equivalent to

$$n \geq \frac{\log\left(\frac{\varepsilon_\theta}{\varepsilon^0}\right)}{\log(q)}.$$

Since the index n has to be a natural number, we are looking for the smallest natural number greater or equal than the right hand side of the previous inequality. ∎

At this point we are provided with all necessary statements to prove the major result of the current section.

Theorem 2.1.12 *Let $\varepsilon = \varepsilon_{min}(\delta, \alpha, h)$ be as defined in Proposition 2.1.6. Further let*

$$\alpha \xrightarrow{\delta \to 0} 0 \qquad \text{and} \qquad \frac{\delta^2}{\alpha} \xrightarrow{\delta \to 0} 0$$

and

$$h = \mathcal{O}(\alpha^\tau) \qquad \alpha \to 0,$$

with $\tau > 1$. Assume $\|y\| > 0$ and $\|y - y^\delta\| \leq \delta$. Define by $N = N(\delta, \alpha, h)$ the index of the first iterate of iteration (2.7) which satisfies

$$\|x^n - x_\alpha^\delta\| \leq \theta \varepsilon,$$

with a constant $\theta > 1$. Then it holds

$$\|x^N - x^\dagger\| \xrightarrow{\delta \to 0} 0.$$

Proof. Note that

$$
\begin{aligned}
\|x^N - x^\dagger\| &\leq \|x^N - x_\alpha^\delta\| + \|x_\alpha^\delta - x^\dagger\| \\
&\leq \theta \varepsilon + \|x_\alpha^\delta - x^\dagger\|,
\end{aligned}
$$

where the first part tends to zero by Lemma 2.1.8 and the second part goes to zero by Theorem 1.5.9. ∎

Before we turn to the question of convergence rates, some remarks arising from the above results should be made.

Remark 2.1.13

1. *We could have used the estimate*

$$\|x_\alpha^\delta\| \leq \frac{1}{(\alpha \omega)^{1/p}} \|y^\delta\|^{2/p} \tag{2.18}$$

instead of $\|x_\alpha^\delta\|$ to define the radii in Lemma 2.1.3 and get computable numbers, but since the estimate contains the factor $\frac{1}{\alpha^{1/p}}$, the radii would have blown up indefinitely for $\alpha \to 0$. This makes it impossible to prove the regularization property in that case. Nevertheless, for fixed α and δ this may give reasonable estimates.

Moreover, it follows from Theorem 1.5.9 that the minimizers x_α^δ are bounded in norm by some positive constant if the regularization parameter was chosen in a proper way. However, this constant is usually unknown.

2. Another important question is, how to notice that some ε-ball is reached or how many iterations are needed to get into this ball?

   To be able to compute the finite index N in Lemma 2.1.11, it is necessary to determine the constant C^1. Therefore we need the initial distance $\varepsilon^0 = \|x^0 - x^\delta_\alpha\|$ as well as the norm of the unknown minimizer x^δ_α. Since this minimizer is unknown, we may estimate ε^0 by some bigger $\tilde{\varepsilon}^0$ and use estimate (2.18) which leads to some different number $\tilde{C}^1 \leq C^1$. Further, the ε_{min}-ball, which depends on the precision h and on α and δ has to be determined via solving the equations in Proposition 2.1.6 substituting $\|x^\delta_\alpha\|$ by estimate (2.18). Afterwards we may choose some target ε_θ greater than ε_{min}. Due to these estimates, we may finally reach at some larger $\tilde{N} \geq N$ which serves as a rather rough estimate for the number of iteration steps necessary to reach the ε_θ-ball. This can be seen as an a priori stopping criterion for iteration (2.7).

3. Most results presented in this section so far have already been published in a slightly different form in [6]. In that paper the numbers R^n and C^n do not depend on the iteration index. Global estimates of these numbers are used instead, which lead to coarser estimates for ε_{min} and the number of necessary iterations N to enter a given ε-ball around the minimizer x^δ_α.

2.1.4   Convergence rates

In the previous paragraph we proved a general regularization result for the adaptive iterated soft-shrinkage method. To assess the quality of the regularization method we will next show how to couple the parameters h, α and δ to achieve certain convergence rates of the regularization scheme. First, we will investigate how to couple the parameters to attain a convergence rate $0 < \kappa < \infty$ for ε, i.e. $\varepsilon = \mathcal{O}(\delta^\kappa)$ with $\delta \to 0$. This coupling will finitely imply convergence rates for Tikhonov-type regularization incorporating sparsity constraints and adaptive operator evaluations.

Lemma 2.1.14 Let the assumptions of Proposition 2.1.6 be valid and assume that $\|y^\delta - y\| \leq \delta$ as well as $\|y\| > 0$. Set

$$\alpha = \mathcal{O}(\delta^\nu) \quad for \quad \delta \to 0,$$

with $0 < \nu < 2$ and

$$h = \mathcal{O}(\alpha^\tau) \quad for \quad \alpha \to 0,$$

with $\tau = 1 + \frac{\kappa}{\nu}$ and $\kappa > 0$. Then it holds

$$\varepsilon = \mathcal{O}(\delta^\kappa) \quad for \quad \delta \to 0.$$

Proof. As we have seen in Lemma 2.1.8, for small α it holds that

$$\varepsilon \sim \frac{h}{\alpha}.$$

By assumption we conclude further

$$
\begin{aligned}
\varepsilon &\sim \alpha^{\tau-1} \\
&\sim \delta^{\nu(\tau-1)} \\
&\sim \delta^{\kappa},
\end{aligned}
$$

which proves the statement.   ∎

This lemma shows us how to couple the parameters α and h to the noise level in such a way that arbitrary convergence rates for ε can be achieved. To obtain convergence rates for the whole regularization scheme, we combine this result with the statements on convergence rates presented in Section 1.5.2. This leads to the following two propositions, which summarize results for different settings.

Proposition 2.1.15 *Let the assumptions of Theorem 1.5.10 and Proposition 2.1.6 be satisfied. Assume further $\|y\| > 0$ and $\|y - y^\delta\| \leq \delta$. Set $\alpha = \frac{\delta^2}{\rho^p}$ and*

$$h = \mathcal{O}(\alpha^\tau) \quad \text{for} \quad \alpha \to 0$$

with $\tau = 1 + \frac{\sigma}{2(\sigma+\mu)}$, where ρ, σ and μ are chosen as defined in Theorem 1.5.10. Further define by $N = N(\delta, \alpha, h)$ the index of the first iterate of iteration (2.7) fulfilling

$$\|x^n - x_\alpha^\delta\| \leq \theta\varepsilon,$$

with a constant $\theta > 1$. Then we obtain

$$\|x^N - x^\dagger\| = \mathcal{O}\left(\delta^{\frac{\sigma}{\sigma+\mu}}\right) \quad \text{for} \quad \delta \to 0.$$

Proof. By having a look at the proof of Lemma 2.1.8 we see that the statement of the lemma also holds in case of $\alpha = \frac{\delta^2}{\rho^p}$ as chosen in Theorem 1.5.10. This implies that Lemma 2.1.14 holds even in case of $\nu = 2$. Together with Theorem 1.5.10 and the triangle inequality, we obtain the suggested statement.   ∎

Proposition 2.1.16 *Let the assumptions of Proposition 2.1.6 and Lemma 2.1.14 be satisfied. Moreover, let $x^\dagger$ satisfy the source condition:*

there is a $\bar{z} \in \bar{H}$ such that $A^*\bar{z} = p \sum_\lambda w_\lambda \operatorname{sgn}(x_\lambda^\dagger)|x_\lambda^\dagger|^{p-1}\varphi_\lambda = \partial\|x^\dagger\|_{\mathbf{w},p}^p.$

Further define by $N = N(\delta, \alpha, h)$ the index of the first iterate of iteration (2.7) fulfilling

$$\|x^n - x_\alpha^\delta\| \leq \theta\varepsilon,$$

with a constant $\theta > 1$.

1. If p and κ are chosen as $1 < p \leq 2$ and $\kappa = \frac{1}{2}$, then it holds

$$\|x^N - x^\dagger\| = \mathcal{O}(\delta^{1/2}) \quad \text{for } \delta \to 0.$$

2. If p and κ are chosen as $1 < p \leq 2$ and $\kappa = \frac{1}{p}$, the operator A obeys the FBI-property and $x^\dagger$ is sparsely represented, it holds

$$\|x^N - x^\dagger\| = \mathcal{O}(\delta^{1/p}) \qquad \text{for } \delta \to 0.$$

Proof. The statements follow directly by combining Lemma 2.1.14 with Theorem 1.5.14 and 1.5.15, respectively.   ∎

The previous propositions finally show, that results on convergence rates which hold in case of exact operator evaluations can be transferred to the adaptive case. This finishes our investigation of the soft-shrinkage case. The case of $p = 1$ is handled briefly in the last part of this chapter. The upcoming section deals with approximate operators satisfying the error estimate (2.3).

2.2   Operator approximations

Even though in the current chapter the focus is on regularization with sparsity constraints combined with adaptive operator evaluations, we will have a brief look at operator approximations. Solving operator equations numerically often forces us to substitute the operator A by an approximate version A_h for example a discretized or compressed version. In the following we assume that an estimate in operator norm of the following form

$$\|A_h - A\| \leq h$$

holds. As mentioned in the introductory chapter, in case of classical Tikhonov regularization, there are some regularization results taking such estimates into account, whereas for non-standard penalty terms such results have not been proved so far.

In the current section we will show that soft-shrinkage in combination with approximated operators is a regularization method as well. Further, it will become clear that the results on convergence rates proved for the exact case carry over to the approximate case.

2.2.1   Basic assumptions and estimates

In the following we will treat again, Tikhonov-type functionals as proposed in [16] with penalty terms as defined in (1.11), i. e.

$$\Gamma_{\alpha\mathbf{w},p,\delta}(x) = \|Ax - y^\delta\|^2 + \alpha\|x\|_{\mathbf{w},p}^p. \tag{2.19}$$

To prove descent and convergence properties for the iterates produced by the soft-shrinkage algorithm, we use again the contraction property of the soft-shrinkage operators. This restricts us to the case $1 < p \leq 2$. Further, we assume that

$$\|A\| < 1 \quad \text{and} \quad \|A_h\| < 1.$$

In the following we consider the shrinkage procedure introduced in Section 1.5.1, but with approximate operators A_h and A_h^* instead of A and A^*, i. e.

$$x^0 \quad \text{arbitrary}$$
$$x^{n+1} = \mathbf{S}_{\alpha\mathbf{w},p}(x^n - A_h^*(A_h x^n - y^\delta)). \tag{2.20}$$

The theory of iterated soft-shrinkage presented in Section 1.5 ensures the convergence of this scheme to the unique minimizer $x_\alpha^{\delta,h}$ of

$$\Gamma_{\alpha\mathbf{w},p,\delta,h}(x) = \|A_h x - y^\delta\|^2 + \alpha\|x\|_{\mathbf{w},p}^p. \tag{2.21}$$

Since we are interested in the minimizer x_α^δ of (2.19), we need to investigate the connection between the minimizers x_α^δ and $x_\alpha^{\delta,h}$. First we prove a technical lemma.

Lemma 2.2.1 *The norm of the minimizer $x_\alpha^{\delta,h}$ of* (2.21) *can be estimated by*

$$\|x_\alpha^{\delta,h}\| \leq \left[\frac{1}{w}\left(\frac{(h\|x^\dagger\|+\delta)^2}{\alpha} + \|x^\dagger\|_{\mathbf{w},p}^p\right)\right]^{1/p} =: M(\alpha,\delta,h,\mathbf{w},p,x^\dagger) = M.$$

Proof. With the help of Lemma 1.5.2 we estimate

$$
\begin{aligned}
\|x_\alpha^{\delta,h}\|^p &\leq \frac{1}{w}\|x_\alpha^{\delta,h}\|_{\mathbf{w},p}^p \\
&\leq \frac{1}{\alpha w}(\|A_h x_\alpha^{\delta,h} - y^\delta\|^2 + \alpha\|x_\alpha^{\delta,h}\|_{\mathbf{w},p}^p) \\
&\leq \frac{1}{\alpha w}(\|A_h x^\dagger - y^\delta\|^2 + \alpha\|x^\dagger\|_{\mathbf{w},p}^p) \\
&\leq \frac{1}{\alpha w}(\|A_h x^\dagger - A x^\dagger + A x^\dagger - y^\delta\|^2 + \alpha\|x^\dagger\|_{\mathbf{w},p}^p) \\
&\leq \frac{1}{w}(\frac{(h\|x^\dagger\|+\delta)^2}{\alpha} + \|x^\dagger\|_{\mathbf{w},p}^p).
\end{aligned}
$$

∎

As a corollary we formulate a well known result on the norm of the minimizer x_α^δ, which follows directly from the previous lemma by taking $h = 0$ or from the proof of Lemma 2.1.8.

Corollary 2.2.2 *For the minimizer x_α^δ of* (2.19) *it holds*

$$\|x_\alpha^\delta\| \leq \left[\frac{1}{w}\left(\frac{\delta^2}{\alpha} + \|x^\dagger\|_{\mathbf{w},p}^p\right)\right]^{1/p} =: \widetilde{M}(\alpha,\delta,\mathbf{w},p,x^\dagger) = \widetilde{M}.$$

The numbers M and $\widetilde{M}$ are constant for fixed error levels δ and h and fixed regularization parameter α. This ensures the convergence of the iterates to the minimizer $x_\alpha^{\delta,h}$ of $\Gamma_{\alpha\mathbf{w},p,\delta,h}$, as we will see in the upcoming lemma. To investigate the regularization property of the considered scheme, we have to make sure that the parameters h, α and δ are coupled in a proper way to keep the constants M and $\widetilde{M}$ finite. This is the topic of the upcoming section.

Remark 2.2.3 *As we have seen in the previous section, there are different ways to estimate the norm of the minimizers of functional* (2.19). *In the current section a formulation incorporating the norm of the true solution $x^\dagger$ has been chosen. This might be estimated by some finite number to get computable constants. Compare Remark 2.1.13 for further comments.*

Next we show that the iterated soft-shrinkage converges linearly. This is a well known result even in case of $p = 1$, see [8]. The proof of the following lemma exploits the contraction property of the shrinkage operators. The result may be also useful to estimate the distance between the iterates x^n produced by iteration (2.20) and the minimizer $x_\alpha^{\delta,h}$ of (2.21) or between the iterates and the minimum-$\|\cdot\|_{\mathbf{w},p}$-solution $x^\dagger$ of $Ax = y$.

Lemma 2.2.4 *The iterated soft-shrinkage algorithm* (2.20) *converges linearly, i. e.*

$$\|x^n - x_\alpha^{\delta,h}\| \leq Dq^n,$$

with $0 < q < 1$ *and a positive constant* $D < \infty$.

Proof. First we show that $x^0 - A_h^*(A_h x^0 - y^\delta)$ and $x_\alpha^{\delta,h} - A_h^*(A_h x_\alpha^{\delta,h} - y^\delta)$ are located in $B(0, R)$ with

$$R = \underbrace{\|x^0 - x_\alpha^{\delta,h}\|}_{=D} + M + \|y^\delta\|.$$

Further we define a local contraction constant for $\mathbf{S}_{\alpha\mathbf{w},p} : B(0, R) \to B(0, R)$ by $q = \frac{1}{1+C}$ with $C = \frac{\alpha\omega p(p-1)}{2} R^{p-2}$, see also Corollary 1.5.5. Then we prove the statement by interlaced induction. We estimate

$$
\begin{aligned}
\|x^0 - A_h^*(A_h x^0 - y^\delta)\| &= \|(I - A_h^* A_h)x^0 - A_h^* y^\delta\| \\
&\leq \|x^0\| + \|y^\delta\| \\
&\leq \|x^0 - x_\alpha^{\delta,h}\| + \|x_\alpha^{\delta,h}\| + \|y^\delta\| \\
&\leq R
\end{aligned}
$$

and

$$
\begin{aligned}
\|x_\alpha^{\delta,h} - A_h^*(A_h x_\alpha^{\delta,h} - y^\delta)\| &\leq \|x_\alpha^{\delta,h}\| + \|y^\delta\| \\
&\leq R.
\end{aligned}
$$

Since $\|x^0 - x_\alpha^{\delta,h}\| \leq Dq^0$, the induction basis is complete.

Next we go for the induction step. With the help of the contraction property of $\mathbf{S}_{\alpha\mathbf{w},p}$ we obtain

$$
\begin{aligned}
&\|x^{n+1} - x_\alpha^{\delta,h}\| \\
&= \|\mathbf{S}_{\alpha\mathbf{w},p}(x^n - A_h^*(A_h x^n - y^\delta)) - \mathbf{S}_{\alpha\mathbf{w},p}(x_\alpha^{\delta,h} - A_h^*(A_h x_\alpha^{\delta,h} - y^\delta))\| \\
&\leq \tfrac{1}{1+C}\|(I - A_h^* A_h)\|\|x^n - x_\alpha^{\delta,h}\| \\
&\leq q\|x^n - x_\alpha^{\delta,h}\| \\
&\leq q^n\|x^0 - x_\alpha^{\delta,h}\|
\end{aligned}
$$

and since $q < 1$ it holds especially that

$$\|x^{n+1} - x_\alpha^{\delta,h}\| \leq D.$$

Further for x^{n+1} it follows

$$
\begin{aligned}
\|x^{n+1} - A_h^*(A_h x^{n+1} - y^\delta)\| &= \|(I - A_h^* A_h)x^{n+1} - A_h^* y^\delta\| \\
&\leq \|x^{n+1}\| + \|y^\delta\| \\
&\leq \|x^{n+1} - x_\alpha^{\delta,h}\| + \|x_\alpha^{\delta,h}\| + \|y^\delta\| \\
&\leq D + M + \|y^\delta\| \\
&= R,
\end{aligned}
$$

which means that $x^{n+1} - A_h^*(A_h x^{n+1} - y^\delta) \in B(0,R)$. Finally, the statement follows by induction. ∎

2.2.2 Regularization property

This paragraph deals with the regularization property of the soft-shrinkage scheme with operator approximations. First we estimate the distance between the minimizer x_α^δ of the functional with respect to the exact operator and the minimizer $x_\alpha^{\delta,h}$ of the functional with respect to the approximated operators.

Proposition 2.2.5 *Let M and $\widetilde{M}$ be as defined in Lemma 2.2.1 and Corollary 2.2.2, then it holds*

$$
\|x_\alpha^{\delta,h} - x_\alpha^\delta\| \leq \tfrac{h}{C}(2\widetilde{M} + \|y^\delta\|),
$$

with $C = \frac{\alpha w p(p-1)}{2}(M + \|y^\delta\|)^{p-2}$.

Proof. First we establish

$$
\|x_\alpha^{\delta,h} - A_h^*(A_h x_\alpha^{\delta,h} - y^\delta)\| \leq \|x_\alpha^{\delta,h}\| + \|y^\delta\|
$$

and

$$
\|x_\alpha^\delta - A^*(A x_\alpha^\delta - y^\delta)\| \leq \|x_\alpha^\delta\| + \|y^\delta\|.
$$

By Lemma 2.2.1 and Corollary 2.2.2 it follows that

$$
\widetilde{M} \leq M
$$

and we define the contraction constant of $\mathbf{S}_{\alpha w, p}$ as $\frac{1}{1+C}$ with $C = \frac{\alpha w p(p-1)}{2}(M + \|y^\delta\|)^{p-2}$ according to Corollary 1.5.5. Further we calculate

$$
\begin{aligned}
\|A_h^* A_h - A^* A\| &= \|A_h^* A_h - A^* A_h + A^* A_h - A^* A\| \\
&\leq \underbrace{\|A_h^* - A^*\|}_{\leq h}\, \underbrace{\|A_h\|}_{<1} + \underbrace{\|A^*\|}_{<1}\, \underbrace{\|A_h - A\|}_{\leq h} \\
&< 2h.
\end{aligned}
$$

Now we can estimate

$$\|x_\alpha^{\delta,h} - x_\alpha^\delta\|$$
$$= \|\mathsf{S}_{\alpha w,p}(x_\alpha^{\delta,h} - A_h^*(A_h x_\alpha^{\delta,h} - y^\delta)) - \mathsf{S}_{\alpha w,p}(x_\alpha^\delta - A^*(A x_\alpha^\delta - y^\delta))\|$$
$$\leq \tfrac{1}{1+C}\|x_\alpha^{\delta,h} - A_h^*(A_h x_\alpha^{\delta,h} - y^\delta) - x_\alpha^\delta + A^*(A x_\alpha^\delta - y^\delta)\|$$
$$= \tfrac{1}{1+C}\|x_\alpha^{\delta,h} - x_\alpha^\delta - A_h^* A_h x_\alpha^{\delta,h} + A_h^* A_h x_\alpha^\delta - A_h^* A_h x_\alpha^\delta + A^* A x_\alpha^\delta + (A_h^* - A^*)y^\delta\|$$
$$\leq \tfrac{1}{1+C}\left(\|I - A_h^* A_h\|\|x_\alpha^{\delta,h} - x_\alpha^\delta\| + \underbrace{\|A_h^* A_h - A^* A\|}_{<2h}\underbrace{\|x_\alpha^\delta\|}_{\leq \widetilde{M}} + \|A_h^* - A^*\|\|y^\delta\| \right)$$
$$\leq \tfrac{1}{1+C}\|x_\alpha^{\delta,h} - x_\alpha^\delta\| + \tfrac{1}{1+C}h(2\widetilde{M} + \|y^\delta\|),$$

which finally leads to

$$\|x_\alpha^{\delta,h} - x_\alpha^\delta\| \leq \tfrac{h}{C}(2\widetilde{M} + \|y^\delta\|). \qquad \blacksquare$$

To formulate a regularization result for the approximate case, we use the same assumption as in the exact case:

Assumption 2.2.6
Pick $\alpha = \alpha(\delta)$ such that

$$\lim_{\delta \to 0} \alpha(\delta) = 0 \qquad and \qquad \lim_{\delta \to 0} \tfrac{\delta^2}{\alpha} = 0.$$

With this assumption and the estimate of the last proposition we are able to state the following regularization result.

Theorem 2.2.7 *Let $A : H \to \bar{H}$ be a bounded, linear operator between separable Hilbert spaces, with $\|A\| < 1$ and let $\|y - y^\delta\| \leq \delta$ as well as $\|y\| > 0$. Assume, that $1 < p \leq 2$, $0 < w < w_\lambda$ for all λ and denote by $x^\dagger$ the unique minimum $\| \cdot \|_{w,p}$-norm solution of $Ax = y$. Let further $\|A_h\| < 1$, Assumption 2.2.6 be satisfied and pick $h = \mathcal{O}(\alpha^\tau)$ for $\alpha \to 0$ with $\tau > 1$. Then it holds that*

$$\lim_{\delta \to 0} \|x_\alpha^{\delta,h} - x^\dagger\| = 0.$$

Proof. First we split up the limit and it follows by Proposition 2.2.5

$$\lim_{\delta \to 0} \|x_\alpha^{\delta,h} - x^\dagger\|$$
$$\leq \lim_{\delta \to 0} \|x_\alpha^{\delta,h} - x_\alpha^\delta\| + \lim_{\delta \to 0} \|x_\alpha^\delta - x^\dagger\|$$
$$\leq \lim_{\delta \to 0} \left(\tfrac{h}{C}(2\widetilde{M} + \|y^\delta\|) \right) + \lim_{\delta \to 0} \|x_\alpha^\delta - x^\dagger\|. \qquad (2.22)$$

The second limit is zero, because of Theorem 1.5.9. So, we only have to justify that the first limit is zero as well. Therefore we have a closer look at the numbers C, M and $\widetilde{M}$. We show that $\tfrac{h}{C}$ tends to zero if δ tends to zero and $2\widetilde{M} + \|y^\delta\|$ is bounded for δ tending to zero.

The number

$$\widetilde{M} = \left[\tfrac{1}{w}\left(\tfrac{\delta^2}{\alpha} + \|x^\dagger\|_{w,p}^p \right) \right]^{1/p}$$

is bounded for $\delta \to 0$, since it holds that $\frac{\delta^2}{\alpha} \to 0$ by Assumption 2.2.6. Further $\|y^\delta\| \leq \delta + \|y\|$ is bounded, which finally implies the boundedness of $2\widetilde{M} + \|y^\delta\|$. Next we investigate

$$M = \left[\frac{1}{w}\left(\frac{(h\|x^\dagger\|+\delta)^2}{\alpha} + \|x^\dagger\|_{\mathbf{w},p}^p\right)\right]^{1/p},$$

which is bounded from below since $\|y\| > 0 \Rightarrow \|x^\dagger\| \neq 0$. This implies, that

$$C = \frac{\alpha w p(p-1)}{2}(M + \|y^\delta\|)^{p-2}$$

behaves like α for $\alpha \to 0$. Again by the choice of $h \sim \alpha^\tau$ we see that $\frac{h}{C} \sim \alpha^{\tau-1}$. This proves that the first limit in (2.22) is zero.   ∎

Remark 2.2.8 *Note that by choosing a proper τ the distance $\|x_\alpha^{\delta,h} - x_\alpha^\delta\|$ may tend to zero in the same way as the distance $\|x_\alpha^\delta - x^\dagger\|$ does. This means that all convergence rate results proved for the exact case carry over to the approximate case.*

The last statement we specify is a corollary of Lemma 2.2.4 and Proposition 2.2.5. It gives a rough estimate for the distances between the iterates of (2.20) and the minimizers x_α^δ of (2.19).

Corollary 2.2.9 *Let the assumptions of Lemma 2.2.4 and Proposition 2.2.5 be satisfied. Then it holds for the iterates generated by iteration (2.20)*

$$\|x^n - x_\alpha^\delta\| \leq D_1 q^n + D_2\frac{h}{C_2},$$

with finite constants $D_1 = \|x^0 - x_\alpha^{\delta,h}\|$ and $D_2 = 2\widetilde{M} + \|y^\delta\|$. The contraction constant $q = \frac{1}{1+C_1}$ is defined by

$$C_1 = \frac{\alpha w p(p-1)}{2}(D_1 + M + \|y^\delta\|)^{p-2}$$

and further, we have

$$C_2 = \frac{\alpha w p(p-1)}{2}(M + \|y^\delta\|)^{p-2}.$$

The numbers M and $\widetilde{M}$ should be as defined in Lemma 2.2.1 and Corollary 2.2.2.

This finishes our investigations on approximate operators, we will now turn back to the case of adaptive operator evaluations. Within the final section of this chapter, a short outlook to the case $p = 1$ is presented.

2.3 Adaptive operator evaluations: $p = 1$

In the current section, we treat the case of adaptive iterated soft-thresholding, i. e. we handle the minimization of Tikhonov-type functionals

$$\Gamma_{\alpha\mathbf{w},1,\delta}(x) = \|Ax - y^\delta\|^2 + \alpha\underbrace{\sum_\lambda w_\lambda|x_\lambda|}_{\Psi_{\alpha\mathbf{w}}(x)}, \tag{2.23}$$

in case of adaptive operator evaluations. Subsequently, we investigate how to choose the noise level h of the adaptive evaluation scheme to ensure that the iterates, calculated via iterated soft-thresholding with adaptive operator evaluations, reach arbitrary small ε-balls around the minimizer x_α^δ of $\Gamma_{\alpha w,1,\delta}$.

Considering exact operator evaluations, iterated soft-thresholding was already investigated, e. g. in [8, 16]. Since the soft-thresholding operator

$$\mathbf{S}_{\alpha w,1}(x) = \sum_\lambda \mathrm{sgn}(x_\lambda)[|x_\lambda| - \tfrac{\alpha w_\lambda}{2}]_+ \varphi_\lambda$$

is not a contraction, the analysis cannot be done in the same way as in the previous sections. Hence, we subsequently follow the lines of [8]. The analysis is based on assumptions on the operator A and uses estimates for the distance of the functional values.

First the general setting of this section is described. We assume again a linear and bounded operator $A : H \rightarrow \bar{H}$ mapping between real, separable Hilbert spaces with $\|A\| < 1$. Moreover, as in the first section of this chapter, we are just provided with approximate operator evaluations $[Ax]_h = Ax + \eta^h$ and $[A^*z]_h = A^*z + \xi^h$ with $\|\eta^h\| \leq h$ and $\|\xi^h\| \leq h$, respectively. Further, we assume that A satisfies the FBI-property, see Definition 1.5.7. This is an essential condition to be able to prove the following statements. It ensures amongst others the uniqueness of the minimizer of (2.23) as mentioned in Remark 1.5.8.

Our first step is again the reformulation of the iteration with approximated operator evaluations

$$x^0 \qquad \text{arbitrary}$$
$$x^{n+1} = \mathbf{S}_{\alpha w,1}(x^n - [A^*([Ax^n]_h - y^\delta)]_h) \tag{2.24}$$

into a version with exact operators. This can be done in the same way as in Section 2.1.1, see Lemma 2.1.1, leading to

$$x^0 \qquad \text{arbitrary}$$
$$x^{n+1} = \mathbf{S}_{\alpha w,1}(x^n - A^*(Ax^n - y^\delta) - \chi^h), \tag{2.25}$$

with $\chi^h = A^*\eta^h + \xi^h$ and $\|\chi^h\| \leq 2h$.

Note that the error terms χ^h are in general different in every iteration step, but since we only have to estimate the norms of these errors, an additional index n is omitted again. To prove descent properties of iteration (2.24), we first need a technical lemma. To simplify notation, we define the distance between the iterates

$$D_n = \|x^n - x^{n+1}\| \tag{2.26}$$

and a Bregman-like distance

$$\Delta_n(x) = \Psi_{\alpha w}(x) - \Psi_{\alpha w}(x^{n+1}) + 2\left\langle A^*(Ax^n - y^\delta), x - x^{n+1}\right\rangle. \tag{2.27}$$

Lemma 2.3.1 Denote by $\{x^n\}_n$ the sequence generated by iteration (2.24) with $\chi^h = A^*\eta^h + \xi^h$ as mentioned above and therefore $\|\chi^h\| \leq 2h$. Let further D_n and Δ_n be as defined in (2.26) and (2.27), then it holds for every $x \in H$ that

$$\Delta_n(x) \geq 2\left\langle x^n - x^{n+1}, x - x^{n+1}\right\rangle - 2\left\langle \chi^h, x - x^{n+1}\right\rangle. \tag{2.28}$$

Further, as long as

$$h < cD_n, \tag{2.29}$$

with $0 < c < \frac{1}{4}$, constants $\bar{c} > 0$ and $\sigma > 0$ exist such that

$$\Delta_n(x^n) \geq \bar{c}D_n^2 \tag{2.30}$$

and

$$\sigma\Delta_n(x^n) \leq \Gamma_{\alpha\mathbf{w},1,\delta}(x^n) - \Gamma_{\alpha\mathbf{w},1,\delta}(x^{n+1}). \tag{2.31}$$

hold true.

Proof. It can be shown that $v = \mathbf{S}_{\alpha\mathbf{w},1}(z)$ solves

$$\min_v \|z - v\|^2 + \Psi_{\alpha\mathbf{w}}(v),$$

see, e. g., [16]. Therefore (2.25) implies

$$0 \in \partial\left(\|x^n - A^*(Ax^n - y^\delta) - \chi^h - \cdot\|^2 + \Psi_{\alpha\mathbf{w}}(\cdot)\right)(x^{n+1}),$$

which is equivalent to

$$2(x^n - A^*(Ax^n - y^\delta) - x^{n+1} - \chi^h) \in \partial\Psi_{\alpha\mathbf{w}}(x^{n+1}).$$

This in turn is equivalent to

$$\Psi_{\alpha\mathbf{w}}(x) \geq \Psi_{\alpha\mathbf{w}}(x^{n+1}) + 2\left\langle x^n - A^*(Ax^n - y^\delta) - x^{n+1} - \chi^h, x - x^{n+1}\right\rangle,$$

for all $x \in H$. By definition of Δ_n this leads to (2.28). Substituting x by x^n and using $\|\chi^h\| \leq 2h$, the Cauchy-Schwarz inequality and the definition of D_n, we get from (2.28)

$$\Delta_n(x^n) \geq 2(D_n^2 - 2hD_n).$$

Inserting the assumption $h < cD_n$, it follows

$$\Delta_n(x^n) \geq (2 - 4c)D_n^2$$

and because of $c < \frac{1}{4}$ the constant $\bar{c} = 2 - 4c$ is positive, which proves (2.30).

To prove (2.31) we show first a general result with the help of Taylor's series expansion, the Cauchy-Schwarz inequality and the assumption $\|A\| < 1$. Let $u, v \in H$ be arbitrarily chosen, then it holds

$$\begin{aligned}
&\Gamma_{\alpha\mathbf{w},1,\delta}(v) - \Gamma_{\alpha\mathbf{w},1,\delta}(u) + \Psi_{\alpha\mathbf{w}}(u) - \Psi_{\alpha\mathbf{w}}(v) + 2\left\langle A^*(Au - y^\delta), u - v\right\rangle \\
=\ & \|Av - y^\delta\|^2 - \|Au - y^\delta\|^2 + 2\left\langle A^*(Au - y^\delta), u - v\right\rangle \\
=\ & 2\int_0^1 \left\langle A^*(A(u + t(v - u)) - y^\delta) - A^*(Au - y^\delta), v - u\right\rangle\ dt \\
=\ & 2\int_0^1 t\left\langle A^*A(v - u), v - u\right\rangle\ dt \\
\leq\ & \|v - u\|^2.
\end{aligned}$$

Inserting x^n and x^{n+1}, the above inequality results in

$$\Gamma_{\alpha\mathbf{w},1,\delta}(x^{n+1}) - \Gamma_{\alpha\mathbf{w},1,\delta}(x^n) + \Delta_n(x^n) \ \leq \ D_n^2$$

and finally with the help of (2.30) this leads to

$$\sigma\Delta_n(x^n) \leq \Gamma_{\alpha\mathbf{w},1,\delta}(x^n) - \Gamma_{\alpha\mathbf{w},1,\delta}(x^{n+1}),$$

where we have set $\sigma = 1 - \frac{1}{\hat{c}}$. As can easily be seen, σ is positive if $c < \frac{1}{4}$ and the proof is complete.   ∎

At this point it should be remarked that the previous lemma holds also in the case of iterated soft-shrinkage, i. e. $1 < p \leq 2$.

The essential condition of Lemma 2.3.1 is (2.29). This condition may be violated after a certain number of iterations, but as long as this condition holds, the functional values at the iterates decrease. Further, we get a decrease of the distance of the iterates to the minimizer x_α^δ as proved in the next proposition. If condition (2.29) is violated, we cannot be sure to get decreasing distances to the minimizer, when continuing the iteration. In this case, it is necessary to choose a smaller noise level h for the adaptive evaluation scheme, in order to satisfy condition (2.29). Later on in this section, a theoretical strategy for choosing admissible levels h to guarantee that the iterates reach arbitrary small ε-balls around the minimizer x_α^δ after a finite number of iterations will be presented.

The next step is to describe the decrease of the distances between the iterates and the minimizer. To this end we define the rest term

$$r_n = \Gamma_{\alpha\mathbf{w},1,\delta}(x^n) - \Gamma_{\alpha\mathbf{w},1,\delta}(x_\alpha^\delta), \tag{2.32}$$

where x_α^δ denotes the unique minimizer of $\Gamma_{\alpha\mathbf{w},1,\delta}$.

Proposition 2.3.2 *Let* $\{x^n\}_n$ *denote a sequence generated by iteration* (2.24). *As long as all assumptions of Lemma 2.3.1 are satisfied and if there exists a constant* $\hat{c} > 0$ *such that*

$$\|x^n - x_\alpha^\delta\|^2 \leq \hat{c}r_n, \tag{2.33}$$

with r_n *as defined in* (2.32), *there exist constants* $\bar{C} > 0$ *and* $\beta \in [0, 1[$ *such that*

$$\|x^n - x_\alpha^\delta\| \leq \bar{C}\beta^n. \tag{2.34}$$

Proof. Note that due to Lemma 2.3.1 the inequality

$$r_n - r_{n+1} = \Gamma_{\alpha\mathbf{w},1,\delta}(x^n) - \Gamma_{\alpha\mathbf{w},1,\delta}(x^{n+1}) \geq \sigma\Delta_n(x^n) \tag{2.35}$$

holds. By convexity and differentiability of $\|A \cdot -y^\delta\|^2$ we have

$$\|Ax_\alpha^\delta - y^\delta\|^2 \geq \|Ax^n - y^\delta\|^2 + 2\left\langle A^*(Ax^n - y^\delta), x_\alpha^\delta - x^n \right\rangle.$$

Combining the last observation with Lemma 2.3.1 and using the Cauchy-Schwarz inequality leads to the following chain of inequalities:

$$
\begin{aligned}
r_n &= \|Ax^n - y^\delta\|^2 + \Psi_{\alpha w}(x^n) - \|Ax_\alpha^\delta - y^\delta\|^2 - \Psi_{\alpha w}(x_\alpha^\delta) \\
&\leq \Psi_{\alpha w}(x^n) - \Psi_{\alpha w}(x_\alpha^\delta) + 2\left\langle A^*(Ax^n - y^\delta), x^n - x_\alpha^\delta \right\rangle \\
&= \Delta_n(x^n) - \Delta_n(x_\alpha^\delta) \\
&\leq \Delta_n(x^n) + 2\left\langle x^n - x^{n+1}, x^{n+1} - x_\alpha^\delta \right\rangle + 2\left\langle \chi^h, x_\alpha^\delta - x^{n+1} \right\rangle \\
&\leq \Delta_n(x^n) + 2\|x^n - x^{n+1}\|\|x^{n+1} - x_\alpha^\delta\| + 4h\|x^{n+1} - x_\alpha^\delta\| \\
&\leq \Delta_n(x^n) + \underbrace{\tfrac{2}{\sqrt{\hat c}}(1 + 2c)}_{c_1} \sqrt{\Delta_n(x^n)}\|x^{n+1} - x_\alpha^\delta\|.
\end{aligned}
$$

Combining this result with (2.35) and applying Young's inequality with $\varepsilon > 0$ results in

$$
\begin{aligned}
\sigma r_n &\leq (r_n - r_{n+1}) + c_1\sqrt{\sigma}\sqrt{\varepsilon}\|x^{n+1} - x_\alpha^\delta\|\,\tfrac{\sqrt{r_n - r_{n+1}}}{\sqrt{\varepsilon}} \\
&\leq (r_n - r_{n+1}) + c_1\left(\tfrac{\sigma\varepsilon}{2}\|x^{n+1} - x_\alpha^\delta\|^2 + \tfrac{1}{2\varepsilon}(r_n - r_{n+1})\right).
\end{aligned}
$$

Using assumption (2.33), (2.35) and setting $\varepsilon = (c_1\hat c)^{-1}$ we get

$$
\sigma r_n \leq (r_n - r_{n+1}) + \tfrac{\sigma}{2}r_n + \tfrac{c_1^2\hat c}{2}(r_n - r_{n+1}).
$$

Rearranging finally leads to

$$
r_{n+1} \leq \left(1 - \tfrac{\sigma(c_1^2\hat c)^{-1}}{2(c_1^2\hat c)^{-1}+1}\right) r_n.
$$

Again, by using (2.33), we see that admissible constants $\beta \in [0, 1[$ and $0 < \bar C$ exist such that (2.34) is fulfilled. ∎

Remark 2.3.3 Note that condition (2.33) might be shown with the help of the FBI property and the boundedness of the evaluations of $\Gamma_{\alpha w, 1, \delta}$ at the iterates, see [8, Lemma 2 and Lemma 3]. The constant $\hat c$ does not depend on the minimization procedure. In general it depends on α, w and δ. In the case of compact operators, this dependence was investigated in [8, Theorem 4].

In the following a strategy to reach a predetermined ε-ball around the minimizer will be presented, i.e.

$$
\|x^N - x_\alpha^\delta\| \leq \varepsilon
$$

with a finite index N. First we prove a short estimate. In the following, we denote by x_U^n the iterates generated by the exact thresholding process (2.5).

Lemma 2.3.4 The distance between the first iterate x_U^1 calculated via the exact thresholding process (2.5) and the first iterate x^1 calculated via the inexact thresholding process (2.24) can be estimated as follows

$$
\|x^1 - x_U^1\| \leq \|\chi^h\|.
$$

Proof. Note that the $S_{\alpha w,1}$ are non-expansive operators, see [16]. This gives us

$$
\begin{aligned}
\|x^1 - x_U^1\| &= \|S_{\alpha w,1}(x^0 - A^*(Ax^0 - y^\delta) - \chi^h) - S_{\alpha w,1}(x^0 - A^*(Ax^0 - y^\delta))\| \\
&\leq \|((x^0 - A^*(Ax^0 - y^\delta)) - \chi^h) - (x^0 - A^*(Ax^0 - y^\delta))\| \\
&= \|\chi^h\|.
\end{aligned}
$$

∎

With fixed parameters α, δ and constant c and provided that we know the constants $\hat{c}$, $\bar{C}$ and β, which are independent of h, we define the following theoretical algorithm:

Algorithm 2.3.5
1. *Fix an arbitrary x^0, $\varepsilon > 0$ and set $n = 0$.*

2. *Set $x_U^0 = x^n$ and set h such that*

$$
h < \frac{\|x_U^0 - x_U^1\|}{c^{-1} + 2}, \tag{2.36}
$$

with $0 < c < \frac{1}{4}$.

3. *Calculate the next iterates via iteration (2.24) as long as*

$$
h < c\|x^n - x^{n+1}\| \tag{2.37}
$$

and

$$
\varepsilon \leq \bar{C}\beta^n.
$$

4. *If*

$$
\bar{C}\beta^n < \varepsilon
$$

stop the iteration. Otherwise, return to step 2.

Remark 2.3.6 *Note that we do not know the distances $\|x_U^0 - x_U^1\|$. But since $x^n = x_U^0 = x_U^1 \Rightarrow x^n = x_\alpha^\delta$, we see that this distance is zero, if and only if the minimum of $\Gamma_{\alpha w,1,\delta}$ is reached. Therefore it is always possible to find a possibly small h satisfying (2.36). To get the proposed iterative process started one may calculate the $(n+1)$th iterate several times via iteration (2.24) for some decreasing noise levels until condition (2.37) is satisfied and hence a suitable noise level h is found.*

Theorem 2.3.7 *Provided that a constant $\hat{c} > 0$ exists such that inequality (2.33) is satisfied for the iterates x^n calculated via Algorithm 2.3.5, then for any $\varepsilon > 0$ there is a finite index N such that*

$$
\|x^N - x_\alpha^\delta\| \leq \varepsilon.
$$

Proof. The iterate $x_U^0 = x^n$ in step 2 of Algorithm 2.3.5 defines a new starting point for the exact iteration (2.5) and the adaptive iteration (2.24). The iterate x^{n+1} can therefore be interpreted as the first iterate of the adaptive iteration with starting point x^n. Choosing h as in step 2 of the algorithm, applying Lemma 2.3.4 and remembering the definition of χ^h gives

$$
\begin{aligned}
c^{-1} h + \|x^{n+1} - x_U^1\| &\leq c^{-1} h + \|\chi^h\| \\
&\leq (c^{-1} + 2) h \\
&< \|x_U^0 - x_U^1\|,
\end{aligned}
$$

which leads to

$$
\begin{aligned}
c^{-1} h &< \|x_U^0 - x_U^1\| - \|x^{n+1} - x_U^1\| \\
&\leq \|x^{n+1} - x_U^0\| \\
&= \|x^n - x^{n+1}\|.
\end{aligned}
$$

Hence, the first condition in step 3 is fulfilled. Further, as long as this condition holds, the required qualifications of Lemma 2.3.1 are satisfied. This especially means that Proposition 2.3.2 can be applied. This proposition gives us a $\bar{C} > 0$ and $\beta \in [0, 1[$ independent of the h, such that

$$
\|x^n - x_\alpha^\delta\| \leq \bar{C} \beta^n.
$$

If we realize, that

$$
\bar{C} \beta^n < \varepsilon
$$

we stop Algorithm 2.3.5. Otherwise we return to step 2 and interpret the n-th iterate as a new starting value. Finally, we have to show that the algorithm stops after a finite number of iterations. This follows, since it is either possible to find a h fulfilling (2.36) or it holds $x_U^0 = x_U^1 = x_\alpha^\delta$. Therefore we reach the ε-ball after at most N iterations, with

$$
N > \frac{\log\left(\frac{\varepsilon}{\bar{C}}\right)}{\log(\beta)}.
$$

∎

Remark 2.3.8 *It should be remarked that the functional values $\Gamma_{\alpha\mathbf{w},1,\delta}(x^n)$ form a decreasing sequence as long as the assumptions of Lemma 2.3.1 are satisfied. This observation might be helpful to prove that inequality (2.33) is satisfied, see also Remark 2.3.3.*

Note that the index N depends on $\bar{C}$ and β, which depend on α and δ, see Proposition 2.3.2. This dependence comes from $\hat{c}$. In general it is not clear how $\hat{c}$ depends on α and δ, see also Remark 2.3.3.

So far we did not reach any regularization results for the case $p = 1$. It seems as if $\hat{c}$ tends to infinity in general, whenever δ tends to zero. Hence, following the approach of this section, it might be impossible to prove the regularization property without further assumptions.

With these investigations we finish our considerations of adaptive iterated soft-shrinkage and turn to the question of how to combine Morozov's discrepancy principle and Tikhonov-type functionals to get an applicable a posteriori parameter strategy for non-standard Tikhonov regularization.

Chapter 3

Morozov's discrepancy principle

This chapter deals with Morozov's discrepancy principle, a widely used a posteriori choice rule for the regularization parameter α. The main idea is to compare the residual $\|Ax_\alpha^\delta - y^\delta\|$ and the noise level δ. Where x_α^δ denotes again a regularized solution of the operator equation

$$Ax = y \tag{3.1}$$

with noisy data y^δ satisfying $\|y - y^\delta\| \leq \delta$ and using some regularization method with regularization parameter α. The ideal parameter α would solve the equation

$$\|Ax_\alpha^\delta - y^\delta\| = \delta, \tag{3.2}$$

since it would not be reasonable to ask for a solution x_α^δ producing a residual with the norm smaller than the noise level. Due to its high non-linearity, this equation is in general not solved directly. There are different approaches to find an approximate solution. One approach was presented in Chapter 1.2, where the discrepancy principle was introduced.

In the following, we use a slightly different formulation. We always look for an α, which fulfills

$$\tau_1\delta \leq \|Ax_\alpha^\delta - y^\delta\| \leq \tau_2\delta, \tag{3.3}$$

with $1 < \tau_1 \leq \tau_2$.

In case of the classical Tikhonov regularization

$$\Gamma_{\alpha,\delta}(x) = \|Ax - y^\delta\|_Y^2 + \alpha\|x\|_X^2,$$

where $A : X \to Y$ is a linear operator between Hilbert spaces, the combination with Morozov's discrepancy principle was studied intensively in the past, see, e.g., [20, 33, 36]. Also in case of nonlinear Operators there are some results for the Hilbert space case, see, e.g., [20, 39].

On the other hand, results for operators mapping between Banach spaces or using penalty terms different from the squared Hilbert space norm are rare. This was the motivation for the work presented in the current chapter. At this point it has to be mentioned that significant parts of the current chapter have already been published in [3]. At almost the same time Jin and Zou had also elaborated some results on this topic, see [30]. Their setting is a bit more general and they focus on the exact solution of equation (3.2), whereas in this thesis we aim at proving results considering the discrepancy principle (3.3).

The main task in the following is to show that Morozov's discrepancy principle combined with generalized Tikhonov-type functionals, like

$$\Gamma_{\alpha,\delta}(x) = \|Ax - y^\delta\|^2 + \alpha\Omega(x) \tag{3.4}$$

with a functional Ω specified later, forms a regularization method. Further, we show convergence rates in terms of Bregman distances assuming some suitable source condition. In case of q-convex functionals Ω we even show convergence rates with respect to norm of the Hilbert space H. For the special case of sparse solutions of the considered operator equation, we show that the convergence rates can be improved. Finally, the second section of the current chapter deals briefly with the case of adaptive operator evaluations in combination with Morozov's discrepancy principle.

3.1   Exact operator evaluations

First we consider the case of exact operator evaluations. We assume a linear operator equation (3.1), where $A : H \to \bar{H}$ is a bounded, linear operator between real separable Hilbert spaces H and $\bar{H}$. Further, we assume that we do not know the exact right hand side, but we are only provided with some noisy data y^δ fulfilling $\|y - y^\delta\| \le \delta$. As a regularization scheme we choose Tikhonov regularization of the form (3.4), where $\Omega : H \to \mathbb{R} \cup \{\infty\}$.

To be able to consider more general penalty terms than the classical square of the Hilbert space norm, we use the concept of Ω-minimizing solutions as defined in Definition 1.2.6 and denote this solution by $x^\dagger$ in the following.

For our investigations in the current section we need the following three conditions.

Condition 3.1.1
Let Ω be

- *strictly convex or convex if A is injective,*

- *non-negative, with $\Omega(x) = 0 \Leftrightarrow x = \mathbf{0}$,*

- *weakly coercive, i.e. $\|x\| \to \infty \Longrightarrow \Omega(x) \to \infty$,*

- *and lower semi-continuous.*

Condition 3.1.2
Let the following implication be satisfied:

$$\Omega(x_n) \to \Omega(x) \Longrightarrow \|x_n\| \to \|x\|,$$

with $\{x_n\}_n \cup \{x\} \subset H$.

Condition 3.1.3
There exists a solution $\bar{x}$ of equation (3.1), which fulfills

$$\Omega(\bar{x}) < \infty.$$

Condition 3.1.1 and 3.1.3 ensure among others the existence and uniqueness of the minimizer of functional (3.4). To be able to show the main regularization result, i. e. $\lim_{\delta \to 0} x_\alpha^\delta = x^\dagger$, we need Condition 3.1.2.

Throughout this section we assume that Ω fulfills the Conditions 3.1.1 and 3.1.3. Condition 3.1.2 is indicated when it is needed. Although these three conditions arise from the proofs below, there are some remarks which should be given at this point for explanation.

Remark 3.1.4

- *For a definition of convexity as well as lower semi-continuity, see Section 1.3.*

- *The first assumption in Condition 3.1.1 ensures the strict convexity of $\Gamma_{\alpha,\delta}$, which is needed to guarantee the uniqueness of its minimizer, see Proposition 3.1.6. To this end it would be sufficient to require Ω to be convex and strictly convex with respect to $N(A)$, i.e. $x_1 - x_2 \in N(A)$, $x_1 \neq x_2$, $t \in\,]0,1[\implies \Omega(tx_1 + (1-t)x_2) < t\Omega(x_1) + (1-t)\Omega(x_2)$.*

- *For the theory presented in the current chapter, the weak lower semi-continuity of Ω is needed, which is equivalent with lower semi-continuity for convex functionals, see Proposition 1.3.4.*

- *To illustrate Condition 3.1.1, we assume a linear, injective and compact operator $K : H \to H$ and consider $\Omega(x) = \|Kx\|^2$. This Ω fulfills the assumptions in Condition 3.1.1 except for the weak coercivity.*

- *In the following we need Ω to satisfy the so-called Kadec property, i.e. $x_n \rightharpoonup x$ and $\Omega(x_n) \to \Omega(x)$ implies $x_n \to x$. As can easily be seen, for weakly convergent sequences Condition 3.1.2 implies the Kadec property.*

To be able to show some of the results below, we need the operator to be weakly continuous. This property is ensured by the following lemma, which is proved, e. g. in [27, Lemma 13.5].

Lemma 3.1.5 *Let X and Y be normed spaces and $A : X \to Y$ be a linear, bounded mapping. Then A is weakly continuous.*

Now we are provided with all necessary assumptions. In the upcoming section we finally prove the main regularization result for the combination of Morozov's discrepancy principle and Tikhonov-type functionals.

3.1.1 Regularization property

Before we will be able to prove the regularization property of the considered method, we have to prove the existence and uniqueness of minimizers as well as the existence of regularization parameters satisfying the discrepancy principle. We start by proving the existence and uniqueness of the minimizer of $\Gamma_{\alpha,\delta}$. We follow the lines of the well-known direct method of the calculus of variations, see, e. g., [2].

Proposition 3.1.6 *Under the global assumptions of this chapter, there is a unique minimizer of functional* (3.4).

Proof. First of all, note that for every $x \in H$ it holds $0 \leq \Gamma_{\alpha,\delta}(x)$. So there is a minimizing sequence $\{x_n\}_n$, i. e.

$$\lim_{n \to \infty} \Gamma_{\alpha,\delta}(x_n) = \inf_{x \in X} \{\Gamma_{\alpha,\delta}(x)\}.$$

Since $\Gamma_{\alpha,\delta}(0) = \|y^\delta\|^2 < \infty$, the infimum is finite. Further, we conclude that the weak coercivity of Ω implies the weak coercivity of $\Gamma_{\alpha,\delta}$. Since $\{\Gamma_{\alpha,\delta}(x_n)\}_n$ is bounded this implies the boundedness of $\{x_n\}_n$. Moreover, H is reflexive and we conclude that there is a weakly convergent subsequence, again denoted by $\{x_n\}_n$, for which holds that $x_n \rightharpoonup \bar{x}$. Exploiting the weak continuity of A and the weak lower semi-continuity of the norm in $\bar{H}$ we obtain the weak lower semi-continuity of $\|A \cdot -y^\delta\|^2$ which together with the weak lower semi-continuity of Ω leads to the weak lower semi-continuity of $\Gamma_{\alpha,\delta}$. This finally leads to

$$\begin{aligned}
\Gamma_{\alpha,\delta}(\bar{x}) &\leq \liminf_{n \to \infty} \Gamma_{\alpha,\delta}(x_n) \\
&= \lim_{n \to \infty} \Gamma_{\alpha,\delta}(x_n) \\
&= \inf_{x \in X} \{\Gamma_{\alpha,\delta}(x)\}.
\end{aligned}$$

This proves the existence of a minimizer.

The uniqueness follows from the strict convexity of $\Gamma_{\alpha,\delta}$.   ∎

Next we have to justify the existence of a unique Ω-minimizing solution for equation (3.1). We state the

Proposition 3.1.7 *Under the global assumptions of this chapter, the operator equation* (3.1) *has a unique Ω-minimizing solution $x^\dagger$.*

Proof. By M we denote the set of solutions of $Ax = y$. By Condition 3.1.3 this set is not empty. We prove the statement by contradiction. Assume there is no Ω-minimizing solution. Then we can find a sequence $\{x_n\}_n$ in M such that $\Omega(x_n) \to c$ with $0 \leq c < \Omega(x)$ for all $x \in M$. The sequence $\{\Omega(x_n)\}_n$ is convergent in $\mathbb{R}$ and hence bounded. Since Ω is weakly coercive, the sequence $\{x_n\}_n$ is also bounded. This implies that we can find a weakly convergent subsequence $\{x_{n_k}\}_k$ converging to some $\hat{x}$. By weak lower semi-continuity of Ω we obtain

$$\begin{aligned}
\Omega(\hat{x}) &\leq \liminf_{k \to \infty} \Omega(x_{n_k}) \\
&= \lim_{k \to \infty} \Omega(x_{n_k}) \\
&= c \\
&< \Omega(x) \quad \text{for all } x \in M.
\end{aligned}$$

Since A is weakly continuous, it follows that

$$x_{n_k} \rightharpoonup \hat{x} \quad \Longrightarrow \quad y = Ax_{n_k} \rightharpoonup A\hat{x},$$

and by the uniqueness of weak limits we conclude $A\hat{x} = y$, which means that $\hat{x} \in M$ and forms a contradiction.

For injective operators A, the solution of (3.1) is unique and therefore also the Ω-minimizing solution. In the case that A is not injective, the uniqueness follows from the strict convexity of Ω. ∎

In the following we will denote by x_α^δ the unique minimizer of (3.4) and the unique Ω-minimizing solution of (3.1) by $x^\dagger$. Further, for fixed noise level δ and $\alpha > 0$ we define the functions

$$m(\alpha) = \|Ax_\alpha^\delta - y^\delta\|^2 + \alpha\Omega(x_\alpha^\delta),$$

$$\Phi(\alpha) = \|Ax_\alpha^\delta - y^\delta\|^2$$

and

$$\Psi(\alpha) = \Omega(x_\alpha^\delta).$$

These functions are single-valued, because of the uniqueness of x_α^δ.

Now we can start moving towards the core statement of this section. The aim is to consider a discrepancy principle of the form (3.3). So the first step is to show that for a given noise level δ there are parameters τ_1, τ_2 and α such that the discrepancy principle is fulfilled. To prove this statement, we state a lemma and a proposition first, which verify some important properties of the functions m, Φ and Ψ.

Lemma 3.1.8 *[51, Chapter II §6] The functions Φ and m are non-decreasing, Ψ is non-increasing on $]0, \infty[$.*

Proof. Choose positive α_1, α_2 arbitrary with $\alpha_1 < \alpha_2$. It holds that

$$m(\alpha_2) = \Phi(\alpha_2) + \alpha_2\Psi(\alpha_2) \geq \Phi(\alpha_2) + \alpha_1\Psi(\alpha_2) \geq \Phi(\alpha_1) + \alpha_1\Psi(\alpha_1) = m(\alpha_1).$$

Therefore m is non-decreasing. Further we have

$$\Phi(\alpha_1) + \alpha_1\Psi(\alpha_1) \leq \Phi(\alpha_2) + \alpha_1\Psi(\alpha_2) \tag{3.5}$$

and

$$\Phi(\alpha_2) + \alpha_2\Psi(\alpha_2) \leq \Phi(\alpha_1) + \alpha_2\Psi(\alpha_1),$$

which implies

$$(\alpha_1 - \alpha_2)\Psi(\alpha_1) \leq (\alpha_1 - \alpha_2)\Psi(\alpha_2).$$

Since $\alpha_1 < \alpha_2$ this proves that Ψ is non-increasing. Finally, by the monotonocity of Ψ and (3.5) it follows that Φ is non-decreasing. ∎

Besides the monotonicity of the functions Φ, Ψ and m, we prove the continuity of these functions. Moreover, we investigate the dependence of the minimizers of the Tikhonov functional on the regularization parameter.

Proposition 3.1.9 *The minimizers x_α^δ depend weakly continuously on α. The functions Φ, Ψ and m are continuous. In case that Condition 3.1.2 is satisfied, the minimizers x_α^δ even depend continuously on α.*

Proof. Let $\{\alpha_n\}_n$ be a positive, arbitrary but fixed sequence, with $\alpha_n \to \bar\alpha > 0$. From the weak coercivity of Ω it follows that $\{x_{\alpha_n}^\delta\}_n$ is bounded. Otherwise it would lead to a contradiction to the minimization property of the $x_{\alpha_n}^\delta$. Since $\{x_{\alpha_n}^\delta\}_n$ is bounded it has a weakly convergent subsequence, again denoted by $\{x_{\alpha_n}^\delta\}_n$ for simplicity, with $x_{\alpha_n}^\delta \rightharpoonup \bar x$. The weak continuity of A, weak lower semi-continuity of the norm and of Ω yields that

$$\|A\bar x - y^\delta\| \le \liminf_{n\to\infty} \|Ax_{\alpha_n}^\delta - y^\delta\| \quad \text{and} \quad \Omega(\bar x) \le \liminf_{n\to\infty} \Omega(x_{\alpha_n}^\delta)$$

and therefore

$$
\begin{aligned}
\|A\bar x - y^\delta\|^2 + \bar\alpha\Omega(\bar x) &\le \liminf_{n\to\infty}(\|Ax_{\alpha_n}^\delta - y^\delta\|^2 + \alpha_n\Omega(x_{\alpha_n}^\delta)) \\
&\le \limsup_{n\to\infty}(\|Ax_{\alpha_n}^\delta - y^\delta\|^2 + \alpha_n\Omega(x_{\alpha_n}^\delta)) \\
&\le \lim_{n\to\infty}(\|Ax - y^\delta\|^2 + \alpha_n\Omega(x)) \\
&= \|Ax - y^\delta\|^2 + \bar\alpha\Omega(x)
\end{aligned}
$$

for all $x \in H$. This implies that $\bar x = x_{\bar\alpha}^\delta$ and that $m(\alpha_n) \to m(\bar\alpha)$. Since the above arguments hold for every weakly convergent subsequence, the whole sequence converges weakly to $x_{\bar\alpha}^\delta$.

Next we show the continuity of Ψ and Φ. We assume that $\Omega(x_{\alpha_n}^\delta) \not\to \Omega(x_{\bar\alpha}^\delta)$. Since Ω is weakly lower semi-continuous, it holds $\Omega(x_{\bar\alpha}^\delta) < \limsup_{n\to\infty} \Omega(x_{\alpha_n}^\delta) = c < \infty$. Further, a subsequence exists, which we denote again by $\{x_{\alpha_n}^\delta\}_n$ such that $\Omega(x_{\alpha_n}^\delta) \to c$. Finally, this leads to

$$
\begin{aligned}
\lim_{n\to\infty} \|Ax_{\alpha_n}^\delta - y^\delta\|^2 &= \lim_{n\to\infty} m(\alpha_n) - \alpha_n\Omega(x_{\alpha_n}^\delta) \\
&= m(\bar\alpha) - \bar\alpha c \\
&= \|Ax_{\bar\alpha}^\delta - y^\delta\|^2 + \bar\alpha(\Omega(x_{\bar\alpha}^\delta) - c) \\
&< \|Ax_{\bar\alpha}^\delta - y^\delta\|^2 \le \liminf_{n\to\infty} \|Ax_{\alpha_n}^\delta - y^\delta\|^2,
\end{aligned}
$$

which forms a contradiction to the weak lower semi-continuity of $\|A \cdot -y^\delta\|^2$. Consequently it holds that $\Omega(x_{\alpha_n}^\delta) \to \Omega(x_{\bar\alpha}^\delta)$ and especially $\Psi(\alpha_n) \to \Psi(\bar\alpha)$. Since $\Phi(\alpha) = m(\alpha) - \alpha\Psi(\alpha)$, we obtain $\Phi(\alpha_n) \to \Phi(\bar\alpha)$.

The strong convergence of the original sequence $\{x_{\alpha_n}^\delta\}_n$ follows with the help of Condition 3.1.2 and Remark 3.1.4. ∎

Now we are able to prove the feasibility of the discrepancy principle (3.3), which follows as a corollary of the theorem below.

Theorem 3.1.10 *Let the global assumptions of this chapter be satisfied and let $\|y - y^\delta\| \le \delta < \|y^\delta\|$, then for every σ with*

$$\|y - y^\delta\| < \sigma < \|y^\delta\|$$

an $\alpha = \alpha(\sigma)$ exists such that

$$\|Ax_\alpha^\delta - y^\delta\| = \sigma. \tag{3.6}$$

Proof. The only thing we have to show is that $I \subset R(\Phi)$, where the interval I is given by $I =]\|y - y^\delta\|^2, \|y^\delta\|^2[$. The statement of the theorem follows then by the monotonicity and continuity of Φ.

Let us first assume an arbitrary sequence $\{\alpha_n\}_n$ with $\alpha_n \to \infty$. Suppose that

$$m(\alpha_n) \longrightarrow \tau > \|y^\delta\|^2.$$

Since

$$\|A\mathbf{0} - y^\delta\|^2 + \alpha\Omega(\mathbf{0}) = \|y^\delta\|^2,$$

holds for all $\alpha > 0$, there must be an $\hat{\alpha} > 0$ with $x^\delta_{\hat{\alpha}} \neq \mathbf{0}$, such that $m(\hat{\alpha}) > \|y^\delta\|^2$, which forms a contradiction to the minimizing property of $x^\delta_{\hat{\alpha}}$. This means that

$$m(\alpha_n) \longrightarrow \tau \leq \|y^\delta\|^2 < \infty \implies \alpha_n\Omega(x^\delta_{\alpha_n}) \leq \tau$$
$$\implies \Omega(x^\delta_{\alpha_n}) \longrightarrow 0. \tag{3.7}$$

Assume that $\{x^\delta_{\alpha_n}\}_n$, with $\alpha_n \to \infty$ but $x^\delta_{\alpha_n} \not\to \mathbf{0}$. By coercivity of Ω this sequence is bounded and we conclude that a weakly convergent subsequence exists denoted by $\{x^\delta_{\alpha_n}\}_n$ as well with

$$x^\delta_{\alpha_n} \rightharpoonup \bar{x} \neq \mathbf{0},$$

then by the weak lower semi-continuity of Ω and $\Omega(x) = 0 \Leftrightarrow x = \mathbf{0}$ we get

$$0 < \Omega(\bar{x}) \leq \liminf_{n\to\infty} \Omega(x^\delta_{\alpha_n}),$$

which contradicts (3.7). Therefore, every weakly convergent subsequence converges weakly to zero, which means that the whole sequence converges weakly to zero. The weak continuity of A and the weak lower semi-continuity of the norm together with (3.7) leads to

$$\|y^\delta\| \leq \liminf_{\alpha_n\to\infty} \|Ax^\delta_{\alpha_n} - y^\delta\| \leq \limsup_{\alpha_n\to\infty} \|Ax^\delta_{\alpha_n} - y^\delta\| \leq \|y^\delta\|$$

and especially $\lim_{\alpha_n\to\infty} \|Ax^\delta_{\alpha_n} - y^\delta\| = \|y^\delta\|$.

Next we assume $\alpha_n \to 0$. For every $\alpha > 0$ we have

$$\begin{aligned} m(\alpha) &= \|Ax^\delta_\alpha - y^\delta\|^2 + \alpha\Omega(x^\delta_\alpha) \\ &\leq \|Ax^\dagger - y^\delta\|^2 + \alpha\Omega(x^\dagger) \\ &= \|y - y^\delta\|^2 + \alpha\Omega(x^\dagger), \end{aligned}$$

which implies $\limsup_{\alpha_n\to0} \|Ax^\delta_{\alpha_n} - y^\delta\| \leq \|y - y^\delta\|$ and completes the proof. ∎

The feasibility of discrepancy principle (3.3) follows immediately.

Corollary 3.1.11 *Under the assumptions of the previous theorem and with $1 < \tau_1 \leq \tau_2 < \frac{\|y^\delta\|}{\delta}$ it exists an $\alpha = \alpha(\delta)$ in order that*

$$\tau_1\delta \leq \|Ax^\delta_\alpha - y^\delta\| \leq \tau_2\delta. \tag{3.8}$$

In our case the assertion of Theorem 3.1.10 can be improved. As proved in [29], equation (3.6) has a unique solution in case that the minimizer of the Tikhonov functional (3.4) is unique for every $\alpha > 0$. So we state and prove this lemma for completeness.

Lemma 3.1.12 *Under the assumptions of Theorem 3.1.10, equation (3.6) has a unique solution.*

Proof. We prove the statement by contradiction. First we assume that there are two minimizer $x_{\alpha_1}^\delta$ and $x_{\alpha_2}^\delta$ such that $\|Ax_{\alpha_1}^\delta - y^\delta\| = \|Ax_{\alpha_2}^\delta - y^\delta\| = \sigma$. Since we know by Proposition 3.1.6 that the minimizer of functional (3.4) is unique for every $\alpha > 0$, this means

$$\|Ax_{\alpha_1}^\delta - y^\delta\|^2 + \alpha_1\Omega(x_{\alpha_1}^\delta) < \|Ax_{\alpha_2}^\delta - y^\delta\|^2 + \alpha_1\Omega(x_{\alpha_2}^\delta)$$

and therefore $\Omega(x_{\alpha_1}^\delta) < \Omega(x_{\alpha_2}^\delta)$. Changing the role of α_1 and α_2 finally leads to a contradiction. ∎

At this point we are almost able to prove the main statement of this section on the regularization property of the combined scheme of Tikhonov funtionals and Morozov's discrepancy principle. So far we only handled the case of a fixed noise level δ. To be able to handle the case of $\delta \to 0$, we need to ensure that the constants τ_1 and τ_2 in (3.8) can be chosen independent of δ at least for small δ. This is the topic of the next lemma.

Lemma 3.1.13 *Let $0 < \|y\|$, $2 < M$ and choose δ^* such that*

$$0 < \delta^* < \frac{1}{M}\|y\|.$$

Assume that for all $0 < \delta \leq \delta^*$ the given data y^δ satisfies $\|y - y^\delta\| \leq \delta$, then it holds

$$1 < M - 1 < \frac{\|y^\delta\|}{\delta} \quad \text{for all } 0 < \delta \leq \delta^*.$$

Proof. Exploiting $\|y - y^\delta\| \leq \delta$ and taking $0 < \delta \leq \delta^* < \frac{1}{M}\|y\|$ into account, we conclude

$$\|y\| - \|y - y^\delta\| \leq \|y^\delta\| \quad \Longrightarrow \quad \frac{\|y\|}{\delta} - \frac{\|y-y^\delta\|}{\delta} \leq \frac{\|y^\delta\|}{\delta}$$

$$\Longrightarrow \quad \frac{\|y\|}{\delta} - 1 \leq \frac{\|y^\delta\|}{\delta}$$

$$\Longrightarrow \quad M - 1 < \frac{\|y^\delta\|}{\delta}.$$

∎

With the help of the previous lemma, we can state the regularization result.

Theorem 3.1.14 *Let the assumptions of Theorem 3.1.10 and Lemma 3.1.13 be fulfilled. Then Corollary 3.1.11 holds for all $0 < \delta \leq \delta^*$ with some $\tau_1 = \tau_1(\delta^*)$, $\tau_2 = \tau_2(\delta^*)$ and δ^* chosen as in Lemma 3.1.13. Hence,*

$$y^\delta \longmapsto T_\alpha(y^\delta) = x_\alpha^\delta = \operatorname*{argmin}_{x \in H} \|Ax - y^\delta\|^2 + \alpha\Omega(x)$$

is a regularization method, if $\alpha = \alpha(\delta)$ is chosen according to the discrepancy principle

$$\tau_1\delta \leq \|Ax_\alpha^\delta - y^\delta\| \leq \tau_2\delta. \tag{3.9}$$

It holds that $x_\alpha^\delta \rightharpoonup x^\dagger$ for $\delta \to 0$, with $x^\dagger$ denoting the unique Ω-minimizing solution of $Ax = y$. Moreover, if Condition 3.1.2 holds, we even get strong convergence to $x^\dagger$.

Proof. We pick $M > 2$ and $\delta^* > 0$ as in Lemma 3.1.13, which means that it is possible to choose τ_1 and τ_2 such that

$$1 < \tau_1 \leq M - 1 \leq \tau_2 < \frac{\|y^\delta\|}{\delta} \quad \text{for all } 0 < \delta \leq \delta^*.$$

This implies that Corollary 3.1.11 is applicable and therefore the parameter choice strategy is realizable.

By the assumptions on τ_1 and y^δ, (3.9) and the minimizing property of x_α^δ we get the following chain of inequalities:

$$
\begin{aligned}
\delta^2 + \alpha\Omega(x_\alpha^\delta) \quad &< \quad \tau_1^2 \delta^2 + \alpha\Omega(x_\alpha^\delta) \\
&\leq \quad \|Ax_\alpha^\delta - y^\delta\|^2 + \alpha\Omega(x_\alpha^\delta) \\
&\leq \quad \|Ax^\dagger - y^\delta\|^2 + \alpha\Omega(x^\dagger) \\
&= \quad \|y - y^\delta\|^2 + \alpha\Omega(x^\dagger) \\
&\leq \quad \delta^2 + \alpha\Omega(x^\dagger).
\end{aligned}
$$

For all $0 < \delta \leq \delta^*$ and $\alpha = \alpha(\delta)$ chosen via (3.9) this implies

$$\Omega(x_\alpha^\delta) < \Omega(x^\dagger) = Q < \infty, \tag{3.10}$$

which means that $x_\alpha^\delta, x^\dagger \in \{x \mid \Omega(x) \leq Q\} = M_Q$. Next we choose a sequence $\{x_{\alpha_n}^{\delta_n}\}_n$ with $\delta_n \to 0$. The weak coercivity of Ω then guarantees that the sequence $\{x_{\alpha_n}^{\delta_n}\}_n$ is bounded, which ensures the existence of a weakly convergent subsequence. Let us denote this subsequence again by $\{x_{\alpha_n}^{\delta_n}\}_n$ and assume $x_{\alpha_n}^{\delta_n} \rightharpoonup \tilde{x}$. For this subsequence it holds

$$
\begin{aligned}
\|A(x_{\alpha_n}^{\delta_n} - x^\dagger)\| \quad &= \quad \|Ax_{\alpha_n}^{\delta_n} - Ax^\dagger\| \\
&= \quad \|Ax_{\alpha_n}^{\delta_n} - y\| \\
&\leq \quad \|Ax_{\alpha_n}^{\delta_n} - y^{\delta_n}\| + \|y^{\delta_n} - y\| \\
&\leq \quad (\tau_2 + 1)\delta_n \longrightarrow 0.
\end{aligned}
\tag{3.11}
$$

The weak continuity of A implies

$$Ax_{\alpha_n}^{\delta_n} \rightharpoonup A\tilde{x}$$

and since weak limits are unique, we conclude $A\tilde{x} = Ax^\dagger = y$. Note that $\tilde{x} - x^\dagger \in N(A)$. The last step is to conclude that $\tilde{x} = x^\dagger$. For injective A we are finished. Assume A to not be injective. We know that $\{x_{\alpha_n}^{\delta_n}\}_n \subset M_Q$ and since Ω is weakly lower semi-continuous we get

$$\Omega(\tilde{x}) \leq \liminf_{n\to\infty} \Omega(x_{\alpha_n}^{\delta_n}) \leq \Omega(x^\dagger).$$

Since $x^\dagger$ is the Ω-minimizing solution, which leads to

$$\Omega(\tilde{x}) = \Omega(x^\dagger)$$

and finally, by Theorem 3.1.7 the Ω-minimizing solution is unique and we have $\tilde{x} = x^\dagger$. Altogether we proved the weak convergence of a subsequence of the original sequence

$\{x^{\delta_n}_{\alpha_n}\}_n$ with $\delta_n \to 0$ to the Ω-minimizing solution. Since by the same arguments, every weak convergent subsequence converges weakly to $x^\dagger$ we conclude that the whole sequence converges weakly to $x^\dagger$.

The last step is to prove strong convergence in the case that Condition 3.1.2 is satisfied. As we have already proved, $x^{\delta_n}_{\alpha_n} \rightharpoonup x^\dagger$. Again, by weak lower semi-continuity of Ω and since $\{x^{\delta_n}_{\alpha_n}\}_n \subset M_Q$ it holds that $\lim_{n\to\infty} \Omega(x^{\delta_n}_{\alpha_n}) = \Omega(x^\dagger)$, which together with Condition 3.1.2 and Remark 3.1.4 proves the strong convergence. ∎

Knowing that the considered combination of Tikhonov regularization and Morozov's discrepancy principle is a regularization method, we turn to the topic of convergence rates in the upcoming section.

3.1.2 Convergence rates

Besides verifying the regularization property of a scheme one is interested in how fast the regularized solutions x^δ_α approach the Ω-minimizing solution $x^\dagger$ when the noise level δ tends to zero. These convergence rates serve as a criterion for comparison of different regularization schemes. To be able to prove convergence rates we have to make additional assumptions on the unknown solution $x^\dagger$. Further, we specialize in particular penalty terms or assume additional properties of the operator to get further improvement of the convergence rates.

A general result

Within this paragraph we suppose that all assumptions we have made in the beginning of Section 3.1 are valid. We use some techniques from Banach space theory to prove a general convergence rate result. Especially, we measure the distance between the regularized solutions x^δ_α and the Ω-minimizing solution $x^\dagger$ in the so-called Bregman distance, see Definition 1.4.9. First, we show convergence rates in Bregman distance, which induce in some cases also convergence rates in norm as we will see in the upcoming paragraphs. For an a priori parameter choice strategy the following result was proved by Burger and Osher, see [11].

To prove this rate for our a posteriori parameter choice rule, it is also necessary to assume the following source condition.

Condition 3.1.15
Let $x^\dagger$ denote the Ω-minimizing solution of (3.1). Assume there is a $\bar{z} \in \bar{H}$ such that the source condition

$$A^* \bar{z} \in \partial\Omega(x^\dagger)$$

is satisfied.

Together with some of the results from the proof of Theorem 3.1.14, we can state a general theorem on the convergence rate of our regularization scheme.

Theorem 3.1.16 Let the assumptions of Theorem 3.1.14 be fulfilled. Further we assume that the Ω-minimizing solution $x^\dagger$ of (3.1) satisfies Condition 3.1.15 and we assume that the regularization parameter α was chosen via the discrepancy principle (3.9). Then for x_α^δ a $d \in \Delta_\Omega(x_\alpha^\delta, x^\dagger)$ exists such that

$$d \leq \|\bar{z}\|(\tau_2 + 1)\delta \quad \text{for all } 0 < \delta \leq \delta^*,$$

with δ^* as defined in Lemma 3.1.13. In case that $\partial\Omega(x^\dagger)$ is single valued, this means

$$\Delta_\Omega(x_\alpha^\delta, x^\dagger) \leq \|\bar{z}\|(\tau_2 + 1)\delta \quad \text{for all } 0 < \delta \leq \delta^*.$$

Proof. From the proof of Theorem 3.1.14, see (3.10) and (3.11), we know

$$\Omega(x_\alpha^\delta) \leq \Omega(x^\dagger)$$

and

$$\|A(x_\alpha^\delta - x^\dagger)\| \leq (\tau_2 + 1)\delta,$$

which together with Condition 3.1.15, gives the estimate

$$\begin{aligned}
d &= \Omega(x_\alpha^\delta) - \Omega(x^\dagger) - \left\langle A^*\bar{z}, x_\alpha^\delta - x^\dagger \right\rangle \ (\in \Delta_\Omega(x_\alpha^\delta, x^\dagger)) \\
&\leq \left\langle \bar{z}, A(x^\dagger - x_\alpha^\delta) \right\rangle \\
&\leq \|\bar{z}\|\|A(x^\dagger - x_\alpha^\delta)\| \\
&\leq \|\bar{z}\|(\tau_2 + 1)\delta.
\end{aligned}$$

The last statement of the theorem is obvious. ∎

Having proved a convergence rate in terms of Bregman distances, we might be interested in cases where the Bregman distance can be estimated from below by the norm of the Hilbert space H. The investigation of this question is the topic of the next paragraph.

To finish this paragraph it is mentioned that Theorem 3.1.16 generalizes the classical situation.

Remark 3.1.17 *In case of $\Omega(\cdot) = \frac{1}{2}\|\cdot\|_H^2$ the Bregman distance of u and v is given by $\Delta_\Omega(u, v) = \frac{1}{2}\|u - v\|_H^2$. Moreover, this means that Theorem 3.1.16 leads to a convergence rate of $\mathcal{O}(\delta^{1/2})$, which is the best possible rate for classical Tikhonov regularization combined with Morozov's discrepancy principle, as remarked in Section 1.2. See also, e. g., [20, Proposition 4.20] for further information.*

Further, if we compare the classical source condition $x^\dagger \in X_\mu = R((A^*A)^{\mu/2})$ with Condition 3.1.15, we note that

$$A^*\bar{z} \in \partial\frac{1}{2}\|x^\dagger\|_H^2 \Leftrightarrow A^*\bar{z} = x^\dagger.$$

This means especially for $\mu = 1$, we get

$$x^\dagger \in X_1 \Rightarrow \text{ Condition 3.1.15 }.$$

The case of q-convex penalty terms

In this paragraph we investigate convergence rates in terms of the Hilbert space norm. Therefore we have to estimate the Bregman distance from below by the norm of the Hilbert space H. In general this is impossible, but we will see that there are many situations where such an estimate is available.

We consider penalty terms $\Omega : H \to \mathbb{R} \cup \{\infty\}$, which are q-convex in some v as defined in Definition 1.4.10, i. e. the inequality

$$c\|u - v\|_H^q \le \Omega(u) - \Omega(v) - \langle v^*, u - v \rangle \tag{3.12}$$

holds for those u with $\|u - v\|_H \le M < \infty$ and with $v^* \in \partial\Omega(v)$. Remembering the Definition 1.4.9 of the Bregman distance, we note that inequality (3.12) can be reformulated as

$$c\|u - v\|_H^q \le d,$$

with $d \in \Delta_\Omega(u, v)$. This observation leads to the following theorem.

Theorem 3.1.18 *Let the assumptions of Theorem 3.1.16 be satisfied. In case there is a $q \in [2, \infty[$ and a finite constant $c > 0$, such that*

$$c\|x_\alpha^\delta - x^\dagger\|_H^q \le d \quad \text{for all} \quad 0 < \delta < \delta^*, \tag{3.13}$$

with $d \in \Delta_\Omega(x_\alpha^\delta, x^\dagger)$ as in Theorem 3.1.16, it holds

$$\|x_\alpha^\delta - x^\dagger\|_H \le C\delta^{1/q} \quad \text{for all} \quad 0 < \delta < \delta^*,$$

with a finite constant $C > 0$.

The proof of this theorem is obvious.

As an example, we turn back to our sparsity enforcing penalty terms of the form

$$\Omega(x) = \|x\|_{\mathbf{w},p}^p = \sum_\lambda w_\lambda |x_\lambda|^p, \tag{3.14}$$

with $1 \le p \le 2$ and $0 < \omega \le w_\lambda$ as introduced in (1.11). As mentioned in Chapter 1.5, Tikhonov regularization utilizing these penalty terms has been studied intensively in recent years and is still a current field of research. Nevertheless, the combination with Morozov's discrepancy principle has been missing so far. We show in the following that Ω as defined in (3.14) fulfills all the assumptions we made in this chapter and therefore lies exactly within the scope of the presented theory.

Lemma 3.1.19 *The functional Ω defined by (3.14) fulfills Condition 3.1.1.*

Proof. The convexity follows, since Ω is a sum of non-negative convex functionals. For $p > 1$ even strict convexity follows. Further, since $\{\varphi_\lambda\}_\lambda$ is a basis of H it holds that $\Omega(x) = 0 \Leftrightarrow x = \mathbf{0}$. The coercivity of Ω follows by Lemma 1.5.2. The lower semi-continuity of Ω follows, since

$$\|x^k - \bar{x}\|_H \to 0 \quad \Rightarrow \quad \sum_\lambda |x_\lambda^k - \bar{x}_\lambda|^2 \to 0 \quad \Rightarrow \quad x_\lambda^k \to \bar{x}_\lambda \quad \text{for all } \lambda,$$

which implies for all $N \in \mathbb{N}$

$$
\begin{aligned}
\sum_{\lambda=1}^{N} w_\lambda |\bar{x}_\lambda|^p &= \sum_{\lambda=1}^{N} \liminf_{k \to \infty} w_\lambda |x_\lambda^k|^p \\
&\leq \liminf_{k \to \infty} \sum_{\lambda=1}^{N} w_\lambda |x_\lambda^k|^p \\
&\leq \liminf_{k \to \infty} \sum_{\lambda} w_\lambda |x_\lambda^k|^p.
\end{aligned}
$$

Finally, the weak lower semi-continuity follows, since Ω is convex and lower semi-continuous. ∎

Lemma 3.1.20 *The functional Ω defined by* (3.14) *fulfills Condition 3.1.2.*

The proof can be found in [16, Lemma 4.3].

To justify that the requirements of Theorem 3.1.18 are fulfilled for $q = 2$, it is referred to a result by Grasmair, Haltmeier and Scherzer.

Lemma 3.1.21 *[24, Lemma 10] Let $1 < p \leq 2$. There exists a constant $c_p > 0$ only depending on p such that*

$$
\frac{c_p}{3\omega + 2\|\bar{u}\|_{\mathbf{w},p}^p + \|u\|_{\mathbf{w},p}^p} \|u - \bar{u}\|^2 \leq \|u\|_{\mathbf{w},p}^p - \|\bar{u}\|_{\mathbf{w},p}^p - \left\langle \partial \|\bar{u}\|_{\mathbf{w},p}^p, u - \bar{u} \right\rangle \tag{3.15}
$$

for all $u, \bar{u} \in D(\| \cdot \|_{\mathbf{w},p}^p)$, for which $\partial \|\bar{u}\|_{\mathbf{w},p}^p \neq \emptyset$.

These lemmas together with the assumption that the operator A fulfills all the necessary properties including the source Condition 3.1.15, ensures that the presented theory is applicable. Theorem 3.1.18 is applicable, since (3.13) holds with $c = c_p/(3(\omega + \|x^\dagger\|_{\mathbf{w},p}^p))$ if we combine (3.15) and (3.10). It provides us with a convergence rate of $\mathcal{O}(\delta^{1/2})$ in case of $1 < p \leq 2$.

In the following, we investigate the case of q-convex penalty terms, which are q-convex with respect to a reflexive Banach space X. We assume a q-convex functional $\bar{\Omega} : X \to \mathbb{R} \cup \{\infty\}$ as defined in Definition 1.4.10, where the reflexive Banach space X is densely and continuously embedded into the Hilbert space H. Due to this embedding property, we obtain

$$
c\|u - v\|_H^q \leq d \tag{3.16}
$$

with some $d \in \Delta_{\bar{\Omega}}(u, v)$. Since the functional $\bar{\Omega}$ is defined on X and not on H, we have to justify that the previous theory presented in this paragraph is also valid in this slightly modified case. We assume that $\bar{\Omega}$ satisfies the Conditions 3.1.1 and 3.1.3 with respect to X. Then we define the functional $\Omega : H \to \mathbb{R} \cup \{\infty\}$ with

$$
\Omega(x) = \begin{cases} \bar{\Omega}(x) & : \quad x \in X \\ \infty & : \quad \text{otherwise} \end{cases} \tag{3.17}
$$

and check whether those conditions are also fulfilled for Ω with respect to H. Note that the minimizers of the considered Tikhonov functional (3.4) will always be elements of X. Therefore we do not need the convexity of Ω to get uniqueness of minimizers. Further, it is obvious that Condition 3.1.3 and the second point in Condition 3.1.1 carry over to Ω. The following lemmas show that also the weak coercivity and the weak lower semi-continuity carry over to Ω.

Lemma 3.1.22 *The functional Ω defined in* (3.17) *is weakly coercive.*

Proof. For every $x \in X$ it holds that $\|x\|_H \leq c\|x\|_X$. Further, we know that $\Omega(x) = \infty$ for all $x \notin X$. Therefore the weak coercivity of $\bar{\Omega}$ implies the weak coercivity of Ω. ∎

Lemma 3.1.23 *The functional Ω defined in* (3.17) *is weakly lower semi-continuous.*

Proof. For every $Q < \infty$ we define the sets

$$
\begin{aligned}
M_Q^X &= \{x \in X \mid \Omega(x) \leq Q\} \qquad \text{and} \\
M_Q^H &= \{x \in H \mid \bar{\Omega}(x) \leq Q\}.
\end{aligned}
$$

Note that these sets coincide, since $\bar{\Omega}(x) = \Omega(x)$, whenever it is finite. According to Definition 1.3.2 we have to show that the sets M_Q^H are weakly closed, i. e. limits of weakly convergent sequences, with respect to H, are elements of M_Q^H.

We pick an arbitrary sequence $\{x_n\}_n \subset M_Q^H$ with $x_n \rightharpoonup x$ in H. Since $\bar{\Omega}$ is weakly coercive, with respect to X it holds that $\{x_n\}_n$ is bounded in X. Therefore there is a subsequence $\{x_{n_k}\}_k$ which is weakly convergent with respect to X and $x_{n_k} \rightharpoonup \bar{x}$. Since M_Q^X is weakly closed with respect to X, we obtain $\Omega(\bar{x}) \leq Q$. Moreover $x_{n_k} \rightharpoonup x$ with respect to H, since it is a subsequence of the original sequence. We know that $X \subset H$ continuously and densely, which leads to

$$
x_{n_k} \overset{X}{\rightharpoonup} \bar{x} \qquad \Longrightarrow \qquad x_{n_k} \overset{H}{\rightharpoonup} \bar{x},
$$

since every element in H^* corresponds to an element in X^* and $\{x_{n_k}\} \subset X$. See also [55, Chapter 23.4] and Remark 3.1.25. By uniqueness of weak limits it follows the statement of the lemma. ∎

These properties of Ω ensure that all statements in Section 3.1.1, which do not require Condition 3.1.2, can be proved in the same way. To prove the statement of Theorem 3.1.16 we assume a slightly different source condition.

Condition 3.1.24
Let $x^\dagger$ denote the Ω-minimizing solution of (3.1). Assume there is a $\bar{z} \in \bar{H}$ such that the source condition

$$
A^* \bar{z} \in \partial \bar{\Omega}(x^\dagger)
$$

is satisfied.

Remark 3.1.25 *Since the embedding of X into H is dense and continuous, it follows $H^* \subset X^*$, i. e.*

$$
\langle y, x \rangle_{X^* \times X} = \langle y, x \rangle_{H^* \times H} \quad \text{for all} \quad x \in X \quad \text{and} \quad y \in H^*,
$$

where $\langle \cdot, \cdot \rangle_{X^* \times X}$ and $\langle \cdot, \cdot \rangle_{H^* \times H}$ denote the dual pairings with respect to X and H, see, e. g., [55, Chapter 23.4]. Further, for every element $x^* \in X^*$ there is a unique extension $\tilde{x}^*$ to H, i. e. $\tilde{x}^* \in H^*$, see, e. g., [1, Chapter 3]. Hence, in Condition 3.1.24 an element $A^* \bar{z}$ has to be understood as an extension of an element in $\partial \bar{\Omega}(x^\dagger)$ to H.

We prove the analogue statement to Theorem 3.1.16 assuming this different source condition.

Theorem 3.1.26 *Let the assumptions of Theorem 3.1.14 be fulfilled. Further, we assume that the Ω-minimizing solution $x^\dagger$ of (3.1) satisfies Condition 3.1.24 and we assume that the regularization parameter α was chosen via the discrepancy principle (3.9). Then for x_α^δ a $d \in \Delta_{\bar{\Omega}}(x_\alpha^\delta, x^\dagger)$ exists such that*

$$d \leq \|\bar{z}\|(\tau_2 + 1)\delta \quad \text{for all } 0 < \delta \leq \delta^*,$$

with δ^* as defined in Lemma 3.1.13. In case that $\partial\bar{\Omega}(x^\dagger)$ is single valued, this means

$$\Delta_{\bar{\Omega}}(x_\alpha^\delta, x^\dagger) \leq \|\bar{z}\|(\tau_2 + 1)\delta \quad \text{for all } 0 < \delta \leq \delta^*.$$

Proof. Note that $x^\dagger$ as well as all x_α^δ are located in X. Further, since the embedding of X into H is dense and continuous, it follows with the help of (3.10), (3.11) and Remark 3.1.25 that

$$
\begin{aligned}
d &= \bar{\Omega}(x_\alpha^\delta) - \bar{\Omega}(x^\dagger) - \left\langle x^*, x_\alpha^\delta - x^\dagger \right\rangle_{X^* \times X} \\
&= \Omega(x_\alpha^\delta) - \Omega(x^\dagger) - \left\langle \tilde{x}^*, x_\alpha^\delta - x^\dagger \right\rangle_{X^* \times X} \\
&\leq -\left\langle \tilde{x}^*, x_\alpha^\delta - x^\dagger \right\rangle_{H^* \times H} \\
&= \left\langle A^* \bar{z}, x^\dagger - x_\alpha^\delta \right\rangle_{H^* \times H} \\
&\leq \left\langle \bar{z}, A(x^\dagger - x_\alpha^\delta) \right\rangle_{H^* \times H} \\
&\leq \|\bar{z}\| \|A(x^\dagger - x_\alpha^\delta)\| \\
&\leq \|\bar{z}\|(\tau_2 + 1)\delta.
\end{aligned}
$$

Here, the element $x^* \in \partial\bar{\Omega}(x^\dagger)$ and $\tilde{x}^*$ denotes its unique extension to H. ■

If we combine the statement of the previous theorem with our observation (3.16), we achieve again a convergence rate of $\mathcal{O}(\delta^{1/q})$, i. e.

$$\|x_\alpha^\delta - x^\dagger\| \leq C\delta^{1/q}.$$

The question is now, which spaces X and functionals Ω fit into the framework of the above situation? In the following we will see that there are indeed some meaningful examples of this setting. First some examples of functionals, which are q-convex in the sense of Definition 1.4.10 are given.

Proposition 3.1.27 *Let X be a q-convex Banach space. Then the functional $\Omega : X \to \mathbb{R}$ defined by*

$$\Omega(x) = \tfrac{1}{p}\|x\|_X^p,$$

with $1 < p \leq q$ is q-convex for every $0 \neq x \in X$.

Proof. Noting that q-convexity of X means

$$\delta_X(\varepsilon) \geq c\varepsilon^q \quad \text{for all } \varepsilon \in]0, 2],$$

where δ_X denotes the modulus of convexity of X, see Definition 1.4.1 and since every q-convex Banach space is also uniformly convex, we can apply Theorem 1.4.7. For all $1 < p < \infty$ we can estimate

$$
\begin{aligned}
\sigma_p(x, y) &= pK_p \int_0^1 \frac{(\max\{\|x-ty\|_X, \|x\|_X\})^p}{t} \delta_X\left(\frac{t\|y\|_X}{2\max\{\|x-ty\|_X, \|x\|_X\}}\right) dt \\
&\geq p\widetilde{K}_p \int_0^1 \frac{(\max\{\|x-ty\|_X, \|x\|_X\})^p}{t} \frac{t^q\|y\|_X^q}{(\max\{\|x-ty\|_X, \|x\|_X\})^q} dt \\
&= p\widetilde{K}_p\|y\|_X^q \int_0^1 t^{q-1}(\max\{\|x-ty\|_X, \|x\|_X\})^{p-q} dt \\
&\geq p\widetilde{K}_p\|y\|_X^q \int_0^1 t^{q-1}(\|x-ty\|_X + \|x\|_X)^{p-q} dt,
\end{aligned}
$$

with some constant $\widetilde{K}_p$ only depending on p. Further, we obtain

$$\|x-y\|_X^p \geq \|x\|_X^p - p\langle j_q(x), y\rangle + p\widetilde{K}_p\|y\|_X^q \int_0^1 t^{q-1}(\|x-ty\|_X + \|x\|_X)^{p-q} dt.$$

Substituting y by $x - y$ and rearrangement leads to

$$K_{p,q,M}\|y-x\|_X^q \leq \tfrac{1}{p}\|y\|_X^p - \tfrac{1}{p}\|x\|_X^p - \langle j_p(x), y-x\rangle,$$

with a constant $K_{p,q,M} = \widetilde{K}_p \int_0^1 t^{q-1}(2\|x\|_X - tM)^{p-q} dt$ depending on p and q, where $\|y-x\|_X \leq M$ with a finite $M > 0$. By assumption we know that $p \leq q$ and $x \neq 0$. This ensures the finiteness of $K_{p,q,M}$ for all y with $\|y-x\|_X \leq M$. Moreover using Theorem 1.4.6 we obtain

$$K_{p,q,M}\|y-x\|_X^q \leq \tfrac{1}{p}\|y\|_X^p - \tfrac{1}{p}\|x\|_X^p - \langle x^*, y-x\rangle,$$

for all y with $\|y-x\|_X \leq M$ and with $x^* \in \partial\tfrac{1}{p}\|x\|_X^p$. This proves the q-convexity of Ω.   ∎

The following example names some function spaces, which are convex of power type and continuously and densely embedded in some Hilbert space.

Example 3.1.28 *For sequence, Lebesgue and Sobolev spaces as mentioned in Example 1.4.4 it holds*

$$\ell_r, L_r(\Omega), W^{m,r}(\Omega) \text{ with } \begin{cases} 1 < r \leq 2 & \text{are} \quad \text{2-convex} \\ 2 \leq r < \infty & \text{are} \quad r\text{-convex} \end{cases}.$$

Further it holds that ℓ_r spaces with $1 < r \leq 2$ are continuously embedded in the Hilbert space ℓ_2 and the Lebesgue spaces $L_r(\Omega)$ with $2 \leq r < \infty$ are continuously embedded in the Hilbert space $L_2(\Omega)$. For the Sobolev spaces it holds by the Sobolev embedding theorem, see, e. g., [1, Theorem 8.9] that

$$m + n(\tfrac{1}{2} - \tfrac{1}{r}) \geq 0, \; m \in \mathbb{N}, \; 1 \leq r < \infty, \; \Omega \in \mathbb{R}^n \Rightarrow W^{m,r}(\Omega) \hookrightarrow L_2(\Omega), \; continuously.$$

To justify the density of an embedding of two spaces it is sufficient to find a function space, which is densely embedded in both of those spaces. Since the space of all finite sequences is densely embedded in every of the above ℓ_r-spaces and the space $C_0^\infty(\Omega)$ is densely embedded in the above $L_r(\Omega)$-spaces, the mentioned continuous embeddings of those spaces are also dense.

The above investigations show that the discrepancy principle can be applied even in case that the penalty term of the Tikhonov functional is not defined on the the Hilbert space H, but only on a continuously and densely embedded reflexive Banach space X. Moreover, if the penalty term is q-convex we are able to prove convergence rates in terms of the Hilbert space H.

In the upcoming paragraph we improve the convergence rates for Tikhonov regularization with Ω as defined in (3.14). Therefore we have to assume among others a sparse structure of the true solution $x^\dagger$.

Convergence rates for sparse solutions

In the current paragraph we handle the case of regularization with sparsity constraints. We make use of the ideas presented by Grasmair, Haltmeier and Scherzer in [24]. They improved convergence rates for Tikhonov regularization with penalty term (3.14) for an a priori choice of the regularization parameter. To achieve this, they assumed the sparsity of the Ω-minimizing solution and the FBI-property of the operator, see Definition 1.5.7. Our aim is to transfer the results from the a priori case to the a posteriori case using the discrepancy principle. Presuming some additional assumptions we are able to show the following improved convergence results.

Theorem 3.1.29 *Let the assumptions of Theorem 3.1.14 and Condition 3.1.15 be satisfied. Let further $1 < p \leq 2$, $x^\dagger$ be sparsely represented with respect to an orthonormal basis $\{\varphi_\lambda\}_\lambda$, A fulfill the FBI-property and α be chosen via discrepancy principle (3.9). Then it holds that*

$$\|x_\alpha^\delta \quad x^\dagger\| = \mathcal{O}(\delta^{1/p}) \quad for \quad \delta \to 0.$$

Proof. First we define by $W^\dagger := \mathrm{span}\{\varphi_\lambda \mid \lambda \in \Lambda^\dagger\}$ a subspace of H, depending on the index set $\Lambda^\dagger = \{\lambda \mid x_\lambda^\dagger \neq 0\}$ and by $\pi_{W^\dagger}$ and $\pi_{W^\dagger}^\perp$ we denote the projections onto $W^\dagger$ and $(W^\dagger)^\perp$ respectively. Using the FBI-property, which implies the existence of some constant $C^\dagger$ such that

$$\|x\| \leq C^\dagger \|Ax\|$$

for all $x \in W^\dagger$, as well as inequality

$$(a + b)^p \leq 2^{p-1}(a^p + b^p) \leq 2(a^p + b^p) \quad \text{for every } a, b > 0 \tag{3.18}$$

and the definition of the projections, we get

$$
\begin{aligned}
\|x_\alpha^\delta - x^\dagger\|^p &= \|\pi_{W^\dagger}(x_\alpha^\delta - x^\dagger) + \underbrace{\pi_{W^\dagger}^\perp(x_\alpha^\delta - x^\dagger)}_{\pi_{W^\dagger}^\perp x_\alpha^\delta}\|^p \\
&\leq 2\|\pi_{W^\dagger}(x_\alpha^\delta - x^\dagger)\|^p + 2\|\pi_{W^\dagger}^\perp x_\alpha^\delta\|^p \\
&\leq 2(C^\dagger)^p \underbrace{\|A\pi_{W^\dagger}(x_\alpha^\delta - x^\dagger)\|^p}_{I} + 2\|\pi_{W^\dagger}^\perp x_\alpha^\delta\|^p.
\end{aligned}
$$

Next we estimate I. Since $\pi_{W^\dagger}^\perp x^\dagger = 0$, by means of the inequality (3.18) and by (3.11), which comes from the proof of Theorem 3.1.14, we get

$$
\begin{aligned}
I &= \|A\pi_{W^\dagger}(x_\alpha^\delta - x^\dagger) + A\pi_{W^\dagger}^\perp x_\alpha^\delta - A\pi_{W^\dagger}^\perp x_\alpha^\delta\|^p \\
&= \|A(x_\alpha^\delta - x^\dagger) + (-A\pi_{W^\dagger}^\perp x_\alpha^\delta)\|^p \\
&\leq 2\|A(x_\alpha^\delta - x^\dagger)\|^p + 2\|A\|^p\|\pi_{W^\dagger}^\perp x_\alpha^\delta\|^p \\
&\leq 2(\tau_2 + 1)^p\delta^p + 2\|A\|^p\|\pi_{W^\dagger}^\perp x_\alpha^\delta\|^p.
\end{aligned}
$$

The final step is to estimate $\|\pi_{W^\dagger}^\perp x_\alpha^\delta\|^p$. Exploiting that the Bregman distances $\Delta_{|\cdot|^p}(x_\lambda, x_\lambda^\dagger)$ are non-negative for all λ, the source Condition 3.1.15, inequality (3.10), which comes also from the proof of Theorem 3.1.14 and again (3.11), we obtain

$$
\begin{aligned}
\|\pi_{W^\dagger}^\perp x_\alpha^\delta\|^p &= \left(\sum_{\lambda \notin \Lambda^\dagger} |(x_\alpha^\delta)_\lambda|^2\right)^{p/2} \\
&\leq \sum_{\lambda \notin \Lambda^\dagger} |(x_\alpha^\delta)_\lambda|^p \\
&\leq \frac{1}{w}\sum_{\lambda \notin \Lambda^\dagger} w_\lambda|(x_\alpha^\delta)_\lambda|^p \\
&= \frac{1}{w}\sum_{\lambda \notin \Lambda^\dagger} w_\lambda[|(x_\alpha^\delta)_\lambda|^p \underbrace{-|x_\lambda^\dagger|^p - p|x_\lambda^\dagger|^{p-1}\operatorname{sgn}(x_\lambda^\dagger)(x_\alpha^\delta - x^\dagger)_\lambda]}_{=0,\ \text{since}\ \lambda \notin \Lambda^\dagger} \\
&\leq \frac{1}{w}\sum_\lambda w_\lambda[|(x_\alpha^\delta)_\lambda|^p - |x_\lambda^\dagger|^p - p|x_\lambda^\dagger|^{p-1}\operatorname{sgn}(x_\lambda^\dagger)(x_\alpha^\delta - x^\dagger)_\lambda] \\
&= \frac{1}{w}\left(\Omega(x_\alpha^\delta) - \Omega(x^\dagger) - \langle x^*, x_\alpha^\delta - x^\dagger\rangle\right) \text{ with } x^* \in \partial\Omega(x^\dagger) \\
&\leq \frac{1}{w}\langle A^*\bar{z}, x^\dagger - x_\alpha^\delta\rangle \\
&\leq \frac{1}{w}\langle \bar{z}, A(x^\dagger - x_\alpha^\delta)\rangle \\
&\leq \frac{1}{w}\|\bar{z}\|\|A(x_\alpha^\delta - x^\dagger)\| \\
&\leq (\tau_2 + 1)\frac{\|\bar{z}\|}{w}\delta.
\end{aligned}
$$

Finally, we arrive at

$$
\begin{aligned}
\|x_\alpha^\delta - x^\dagger\|^p &\leq 4(C^\dagger)^p(\tau_2 + 1)^p\delta^p + \left(2(C^\dagger)^p\|A\|^p + 1\right)2(\tau_1 + 1)\frac{\|\bar{z}\|}{w}\delta \\
&\leq C_1^\dagger\delta^p + C_2^\dagger\delta.
\end{aligned}
$$

This implies

$$
\|x_\alpha^\delta - x^\dagger\| = \mathcal{O}(\delta^{1/p}) \quad \text{for} \quad \delta \to 0. \qquad \blacksquare
$$

Unfortunately, the case $p = 1$ cannot be handled in the same way. We deal with this case in the upcoming theorem.

Theorem 3.1.30 *Let the assumptions of Theorem 3.1.14 and Condition 3.1.15 be satisfied. Let further $p = 1$, A fulfill the FBI-property and α be chosen via discrepancy principle* (3.9). *Then it holds that*

$$\|x_\alpha^\delta - x^\dagger\| = \mathcal{O}(\delta) \quad for \quad \delta \to 0.$$

Proof. We define an index set $\Lambda_\omega = \{\lambda \mid |x_\lambda^*| \geq \omega\}$ with $x^* = A^*\bar{z}$. Further we define a subspace $W_\omega = \operatorname{span}\{\varphi_\lambda \mid \lambda \in \Lambda_\omega\}$ and the projections π_{W_ω} and $\pi_{W_\omega}^\perp$ onto W_ω and $(W_\omega)^\perp$ respectively. Since $x^* \in H$, it follows that $|\Lambda_\omega| < \infty$, see Lemma 1.5.13, and therefore W_ω is finite dimensional. Exploiting the FBI-property implies then

$$\|x\| \leq C_\omega \|Ax\| \quad \text{for all} \quad x \in W_\omega$$

with a finite constant C_ω. Moreover, by $x^* \in \partial\|x^\dagger\|_{\mathbf{w},1}$ we conclude for all $\lambda \notin \Lambda_\omega$ that $x_\lambda^\dagger = 0$. Taking these observations into account and using (3.11), we obtain

$$
\begin{aligned}
&\|x_\alpha^\delta - x^\dagger\| \\
\leq{}& \|\pi_{W_\omega}(x_\alpha^\delta - x^\dagger)\| + \|\pi_{W_\omega}^\perp(x_\alpha^\delta - x^\dagger)\| \\
\leq{}& C_\omega\|A\pi_{W_\omega}(x_\alpha^\delta - x^\dagger) + A\pi_{W_\omega}^\perp(x_\alpha^\delta - x^\dagger) - A\pi_{W_\omega}^\perp(x_\alpha^\delta)\| + \|\pi_{W_\omega}^\perp(x_\alpha^\delta)\| \\
\leq{}& C_\omega\|A(x_\alpha^\delta - x^\dagger)\| + (C_\omega\|A\| + 1)\|\pi_{W_\omega}^\perp(x_\alpha^\delta)\| \\
\leq{}& C_\omega(\tau_2 + 1)\delta + (C_\omega\|A\| + 1)\|\pi_{W_\omega}^\perp(x_\alpha^\delta)\|.
\end{aligned}
$$

Finally, we estimate $\|\pi_{W_\omega}^\perp(x_\alpha^\delta)\|$. Note that $m = \max\{|x_\lambda^*| \mid \lambda \notin \Lambda_\omega\}$ is well-defined, since $\{x_\lambda^*\}_\lambda \in \ell_2$. Further we have that $0 \leq m < \omega$ and $x_\lambda^* \leq m$ for all $\lambda \notin \Lambda_\omega$. In addition to these observations we exploit $x^* \in \partial\|x^\dagger\|_{\mathbf{w},1}$, (3.10) and (3.11) and obtain

$$
\begin{aligned}
\|\pi_{W_\omega}^\perp(x_\alpha^\delta)\| &= \left(\sum_{\lambda\notin\Lambda_\omega} |(x_\alpha^\delta)_\lambda|^2\right)^{1/2} \\
&\leq \sum_{\lambda\notin\Lambda_\omega} |(x_\alpha^\delta)_\lambda| \\
&\leq \frac{1}{\omega-m}\sum_{\lambda\notin\Lambda_\omega}(w_\lambda - m)|(x_\alpha^\delta)_\lambda| \\
&\leq \frac{1}{\omega-m}\sum_{\lambda\notin\Lambda_\omega}(w_\lambda|(x_\alpha^\delta)_\lambda| - x_\lambda^*(x_\alpha^\delta)_\lambda) \\
&= \frac{1}{\omega-m}\sum_{\lambda\notin\Lambda_\omega}(w_\lambda|(x_\alpha^\delta)_\lambda| - w_\lambda|x_\lambda^\dagger| - x_\lambda^*(x_\alpha^\delta - x^\dagger)_\lambda) \\
&\leq \frac{1}{\omega-m}\sum_{\lambda}(w_\lambda|(x_\alpha^\delta)_\lambda| - w_\lambda|x_\lambda^\dagger| - x_\lambda^*(x_\alpha^\delta - x^\dagger)_\lambda) \\
&= \frac{1}{\omega-m}\left(\|x_\alpha^\delta\|_{\mathbf{w},1} - \|x^\dagger\|_{\mathbf{w},1} - \langle x^*, x_\alpha^\delta - x^\dagger\rangle\right) \\
&\leq \frac{1}{\omega-m}\langle A\bar{z}, x^\dagger - x_\alpha^\delta\rangle \\
&\leq \frac{1}{\omega-m}\|\bar{z}\|\|A(x^\dagger - x_\alpha^\delta)\| \\
&\leq \frac{1}{\omega-m}\|\bar{z}\|(\tau_2 + 1)\delta.
\end{aligned}
$$

We arrive at

$$\|x_\alpha^\delta - x^\dagger\| \le \left(C_\omega + (C_\omega\|A\| + 1)\tfrac{1}{\omega-m}\|\bar{z}\|\right)(\tau_2 + 1)\delta,$$

which proves the statement.   ∎

Remark 3.1.31 *Assuming $p = 1$ and $x^\dagger \ne 0$ the source Condition 3.1.15 reads as: a $\bar{z} \in H$ exists such that*

$$
\begin{aligned}
A^*\bar{z} &\in \partial[\sum_\lambda w_\lambda |x_\lambda^\dagger|] \\
&= \sum_\lambda w_\lambda \operatorname{sgn}(x_\lambda^\dagger)\varphi_\lambda.
\end{aligned}
$$

This means especially that it holds, at least for all $\lambda \in \Lambda^\dagger$, that

$$\varphi_\lambda \in R(A^*).$$

We finish our investigations of the discrepancy principle for the case of exact operator evaluations and turn briefly to the case of inexact evaluations. Therefore we make use of the results presented in Chapter 2.

3.2   Adaptive operator evaluations

In this section we handle the special case of regularization with sparsity constraints combined with Morozov's discrepancy principle and adaptive operator evaluations. The functional which we want to minimize is again the Tikhonov-type functional (3.4) with sparsity enforcing penalty term as in (3.14). We use some results from Chapter 2 and combine them with results from the current chapter.

To begin with we show some technical statements. The first statement proves that α chosen via discrepancy principle (3.9) incorporating the exact operator A is not zero as long as δ is strictly positive.

Lemma 3.2.1 *Let the assumptions of Theorem 3.1.14 be satisfied. Then α chosen via the discrepancy principle (3.9) is positive as long as $\delta > 0$.*

Proof. Assume that $\delta > 0$ but $\alpha = 0$. By the discrepancy principle and the proof of Theorem 3.1.10 we obtain

$$
\begin{aligned}
(\tau_1)^2\delta^2 &\le \|Ax_\alpha^\delta - y^\delta\|^2 \\
&\le \|y - y^\delta\|^2 \\
&\le \delta^2.
\end{aligned}
$$

This chain of inequalities forms a contradiction, since $\tau_1 > 1$.   ∎

Next we show that the quotient $\frac{\delta^2}{\alpha}$ is bounded if α was chosen via Morozov's discrepancy principle with exact operator A.

Lemma 3.2.2 *Assume that τ_1 was chosen as in Theorem 3.1.14 and $\alpha = \alpha(\delta)$ was chosen via the discrepancy principle* (3.9). *Then it holds for all $0 < \delta \leq \delta^*$ that*

$$\frac{\delta^2}{\alpha} \leq \frac{1}{(\tau_1)^2 - 1} \|x^\dagger\|_{\mathbf{w},p}^p.$$

Proof. Exploiting the discrepancy principle, the minimizing property of x_α^δ and by the statement of the previous lemma we obtain the following chain of inequalities

$$\begin{aligned}
\frac{(\tau_1)^2 \delta^2}{\alpha} &\leq \frac{1}{\alpha} \|Ax_\alpha^\delta - y^\delta\|^2 \\
&\leq \frac{1}{\alpha} (\|Ax_\alpha^\delta - y^\delta\|^2 + \alpha \|x_\alpha^\delta\|_{\mathbf{w},p}^p) \\
&\leq \frac{1}{\alpha} (\|Ax^\dagger - y^\delta\|^2 + \alpha \|x^\dagger\|_{\mathbf{w},p}^p) \\
&\leq \frac{1}{\alpha} (\|y - y^\delta\|^2 + \alpha \|x^\dagger\|_{\mathbf{w},p}^p) \\
&\leq \frac{\delta^2}{\alpha} + \|x^\dagger\|_{\mathbf{w},p}^p.
\end{aligned}$$

This is equivalent to

$$\frac{\delta^2}{\alpha} \leq \frac{1}{(\tau_1)^2 - 1} \|x^\dagger\|_{\mathbf{w},p}^p. \qquad \blacksquare$$

The third technical statement is the boundedness of the $\|x_\alpha^\delta\|$, which follows directly from the proof of Theorem 3.1.14.

Lemma 3.2.3 *Assume that the requirements of the previous lemma are satisfied. Then it holds*

$$\|x_\alpha^\delta\| \leq \frac{1}{\omega^{1/p}} \|x^\dagger\|_{\mathbf{w},p}$$

for all $0 < \delta \leq \delta^*$.

Proof. By Lemma 1.5.2 and (3.10) we obtain

$$\|x_\alpha^\delta\|^p \leq \frac{1}{\omega} \|x_\alpha^\delta\|_{\mathbf{w},p}^p < \frac{1}{\omega} \|x^\dagger\|_{\mathbf{w},p}^p$$

which proves the statement. $\qquad \blacksquare$

Assuming that $\|x^\dagger\|_{\mathbf{w},p} < \rho$ for some finite $\rho > 0$, we can redefine the radius R^n in Lemma 2.1.3, as

$$R^n = \varepsilon^n + \frac{1}{\omega^{1/p}} \rho + \|y^\delta\| + 2h.$$

The analysis remains the same, but in Proposition 2.1.6 we get a different equation for determining $\varepsilon_{\min}$:

$$\varepsilon^{\frac{p-1}{p-2}} + \left[\frac{1}{\omega^{1/p}} \rho + \|y^\delta\| + 2h \right] \varepsilon^{\frac{1}{p-2}} - \left(\frac{2}{\omega p (p-1)} \right)^{\frac{1}{p-2}} \left(\frac{h}{\alpha} \right)^{\frac{1}{p-2}} = 0.$$

This equation is equivalent to

$$\left[\varepsilon + \frac{1}{\omega^{1/p}} \rho + \|y^\delta\| + 2h \right] \varepsilon^{\frac{1}{p-2}} = \left(\frac{2}{\omega p (p-1)} \right)^{\frac{1}{p-2}} \left(\frac{h}{\alpha} \right)^{\frac{1}{p-2}}.$$

Following the lines of the proof of Lemma 2.1.8 we conclude, since for $h, \delta, \varepsilon \to 0$ the factor $\varepsilon + \frac{1}{\omega^{1/p}} \rho + \|y^\delta\| + 2h$ is bounded and hence

$$\varepsilon_{\min} = \mathcal{O}(\tfrac{h}{\alpha}).$$

Especially, there is a constant c such that $\varepsilon_{\min} \leq c\frac{h}{\alpha}$ for all $0 < \delta \leq \delta^*$. Finally, we get for some given $\gamma > 0$, $\theta > 1$, $h = \beta\delta^3$ and $\beta > 0$ chosen small enough that

$$
\begin{aligned}
\varepsilon + h &= \theta\varepsilon_{\min} + h \\
&\leq \theta c\frac{h}{\alpha} + h \\
&\leq \theta c\beta\frac{\delta^2}{\alpha}\delta + \beta\delta^3 \\
&\leq \left(\theta c\frac{\delta^2}{\alpha} + \delta^2\right)\beta\delta \\
&\leq \gamma\delta.
\end{aligned}
\tag{3.19}
$$

The existence of such a β follows from Lemma 3.2.2.

In practice it might be difficult to choose an appropriate β to meet the assumption of the following theorem, since γ will in general be quite small. Nevertheless, we state the upcoming theorem which gives us a theoretical feeling for the combination of Morozov's discrepancy principle and adaptive iterated soft-shrinkage as discussed in Chapter 2.

Theorem 3.2.4 Let $\|y - y^\delta\| \leq \delta \leq \|y^\delta\|$ for $0 < \delta \leq \delta^*$ and let $\|A\| < 1$, $1 < p \leq 2$ and further $\tau_1 = \tau_1(\delta^*)$, $\tau_2 = \tau_2(\delta^*)$, $\gamma > 0$ such that

$$1 < (\tau_1 - \gamma) < \tau_1 \leq \tau_2 < (\tau_2 + \gamma) < \frac{\|y^\delta\|}{\delta}, \quad \text{for all } 0 < \delta \leq \delta^*.$$

Choose $\theta > 1$ and $h = \beta\delta^3$ with $\beta > 0$ small enough, then an $\alpha = \alpha(\delta)$ exists such that

$$(\tau_1 - \gamma)\delta \leq \| \left[Ax^N \right]_h - y^\delta \| \leq (\tau_2 + \gamma)\delta \tag{3.20}$$

holds, with $N = N(\delta, \alpha, h)$ such that

$$\|x^N - x^\delta_\alpha\| \leq \varepsilon = \theta\varepsilon_{\min}. \tag{3.21}$$

The parameter $\varepsilon_{\min} = \varepsilon_{\min}(\delta, \alpha, h)$ is given by

$$\varepsilon_{\min} = \frac{2h}{\alpha\omega},$$

for $p = 2$ and for $1 < p < 2$ as the solution of

$$\varepsilon^{\frac{p-1}{p-2}} + \left[\frac{1}{\omega^{1/p}}\rho + \|y^\delta\| + 2h\right]\varepsilon^{\frac{1}{p-2}} - \left(\frac{2}{\omega p(p-1)}\right)^{\frac{1}{p-2}}\left(\frac{h}{\alpha}\right)^{\frac{1}{p-2}} = 0.$$

Further, it holds then

$$x^N \xrightarrow{\delta \to 0} x^\dagger.$$

Proof. First we have to show that there is an $\tilde{\alpha}$ such that the stated discrepancy principle holds. Considering the exact case, assume that the discrepancy principle

$$\tau_1\delta \leq \|Ax^\delta_\alpha - y^\delta\| \leq \tau_2\delta$$

holds for $\tilde{\alpha}$. As we have seen in (3.19), with $h = \beta\delta^3$ and β small enough, we obtain

$$\varepsilon + h \leq \gamma\delta.$$

Further using (3.21) we can estimate

$$
\begin{aligned}
\left\| \left[Ax^N \right]_h - y^\delta \right\| &= \left\| Ax^N + \xi^{h,N} - Ax_{\tilde{\alpha}}^\delta + Ax_{\tilde{\alpha}}^\delta - y^\delta \right\| \\
&\leq \left\| x^N - x_{\tilde{\alpha}}^\delta \right\| + h + \left\| Ax_{\tilde{\alpha}}^\delta - y^\delta \right\| \\
&\leq \varepsilon + h + \left\| Ax_{\tilde{\alpha}}^\delta - y^\delta \right\| \\
&\leq (\gamma + \tau_2)\delta,
\end{aligned}
$$

with the notation $\left[Ax^N \right]_h = Ax^N + \xi^{h,N}$ from Chapter 2. On the other hand we get

$$
\begin{aligned}
\tau_1\delta &\leq \left\| Ax_{\tilde{\alpha}}^\delta - y^\delta \right\| \\
&= \left\| Ax_{\tilde{\alpha}}^\delta - \left[Ax^N \right]_h + \left[Ax^N \right]_h - y^\delta \right\| \\
&\leq \left\| A(x_{\tilde{\alpha}}^\delta - x^N) - \xi^{h,N} + \left[Ax^N \right]_h - y^\delta \right\| \\
&\leq \left\| x_{\tilde{\alpha}}^\delta - x^N \right\| + h + \left\| \left[Ax^N \right]_h - y^\delta \right\| \\
&\leq \gamma\delta + \left\| \left[Ax^N \right]_h - y^\delta \right\|
\end{aligned}
$$

or equivalently

$$(\tau_1 - \gamma)\delta \leq \left\| \left[Ax^N \right]_h - y^\delta \right\|.$$

This means (3.20) is satisfied if we choose $\alpha = \tilde{\alpha}$.

Finally, we show the convergence result. From the exact case we know

$$\left\| x_\alpha^\delta - x^\dagger \right\| \xrightarrow{\delta \to 0} 0.$$

Further for small δ it holds by Lemma 3.2.2, (3.21) and since $h = \beta\delta^3$ that

$$\left\| x^N - x_\alpha^\delta \right\| \leq c\delta$$

with some finite constant $c > 0$ independent of δ. Via the triangle inequality we get

$$\left\| x^N - x^\dagger \right\| \leq \left\| x^N - x_\alpha^\delta \right\| + \left\| x_\alpha^\delta - x^\dagger \right\| \xrightarrow{\delta \to 0} 0. \qquad \blacksquare$$

The previous theorem shows that the discrepancy principle can be applied in principle. The existence of a regularization parameter α satisfying (3.20) and (3.21) is guaranteed, but it might be difficult to determine such an α in practice. The strategy may go as follows: First pick an α_0. Calculate x^N such that (3.21) is satisfied and check whether or not (3.20) holds also. If it does not hold decrease α for example by picking $\alpha_n = q^n\alpha_0$ with $0 < q < 1$.

Chapter 4

Numerical investigations

Within this chapter we consider an inverse heat conduction problem from steel production and solve it by a combination of iterated soft-shrinkage and an adaptive finite element method. First the inverse problem will be introduced and afterwards the implementation and some numerical calculations will be discussed.

4.1   An inverse heat conduction problem

The considered problem is motivated by monitoring steel production. When steel is produced, iron ore has to be melted in steel furnaces. These furnaces are brought on stream once and stay in operation for several years. Since they are heated throughout that whole period, it is difficult to measure the temperature inside the furnace. Every measurement device installed inside the furnace would be destroyed due to high temperatures. Nevertheless, it is important to know about the shape of the inner boundary of the furnace. This is because physical and chemical processes cause damage of the furnace wall. If the wall thickness falls below a critical value it might break through. To avoid this breakdown, a steel furnace has to be turned off at certain times, to check the state of the furnace walls.

If it would be possible to determine the shape of the inner boundary by indirect measurements outside the furnace, one could run the furnace without a break until the critical thickness of the furnace wall is reached. One possibility to get information on the shape of the inner boundary is to determine the temperature at a surface lying entirely within the furnace wall near the inner boundary. Since we expect that some parts of the wall are damaged more than others, we should be able to detect some hotspots in the reconstructed temperature profile indicating these damaged parts of the wall.

Mathematically we model this problem as a two dimensional inverse heat conduction problem (IHCP) on a ring-shaped domain Ω, see Figure 4.1. We define $\Omega = \{x \in \mathbb{R}^2 \mid 0 < r_0 < \|x\| < r_1\}$, the inner boundary $\Gamma_0 = \{x \in \mathbb{R}^2 \mid \|x\| = r_0\}$ and the outer boundary $\Gamma_1 = \{x \in \mathbb{R}^2 \mid \|x\| = r_1\}$. The direct problem is given by the following heat equation:

$$
\begin{aligned}
\frac{\partial u}{\partial t} - \kappa \Delta u &= 0 && \text{in } (0, T) \times \Omega \\
u &= f && \text{on } (0, T) \times \Gamma_0 \\
-\kappa \frac{\partial u}{\partial \nu} &= k && \text{on } (0, T) \times \Gamma_1 \\
u(0, \cdot) &= u_0 && \text{in } \Omega.
\end{aligned}
\tag{4.1}
$$

Figure 4.1: The ring-shaped domain Ω.

Since the inner boundary Γ_0 is located entirely within the furnace wall, we can assume a constant heat conductivity κ. Further, we assume that the heat flux through the outer boundary Γ_1 and the temperature at this boundary can be measured during the whole time interval $(0, T)$. The latter measurements serve as additional information to formulate the inverse problem.

Next we reformulate the forward problem (4.1) as an operator equation. Let

$$L(f; k, u_0) = u$$

be the solution operator of problem (4.1), i. e. a non-linear operator, which maps some boundary function f on the inner boundary to the solution of (4.1) with fixed initial value u_0 and Neumann boundary value k at the outer boundary. To be able to deal with a linear operator equation, we split up the operator into a linear and a constant part

$$
\begin{aligned}
L(f; k, u_0) &= L(f; 0, 0) + L(0; k, u_0) \\
&= u_l + u_c
\end{aligned}
$$

and consider only the first part in the following investigations. In other words, we assume that the second part u_c has been subtracted. For simplicity, we set further $\kappa = 1$ and $T = 1$. Therefore we obtain a simplified heat equation

$$
\begin{array}{ll}
\frac{\partial u}{\partial t} - \Delta u = 0 & \text{in } (0,1) \times \Omega \\
u = f & \text{on } (0,1) \times \Gamma_0 \\
-\frac{\partial u}{\partial \nu} = 0 & \text{on } (0,1) \times \Gamma_1 \\
u(0, \cdot) = 0 & \text{in } \Omega,
\end{array}
\tag{4.2}
$$

which is equivalent to a linear operator equation

$$Lf = u.$$

Finally, to be able to formulate the inverse problem incorporating the measured temperature at the outer boundary, we restrict the solution of (4.2) to the outer boundary and obtain the linear operator equation describing the direct problem

$$Lf|_{(0,1) \times \Gamma_1} = Kf = g.$$

The inverse problem is now to reconstruct the boundary data f from measurement data g on the outer boundary. These data are usually perturbed. In the following we assume temperature measurements g^δ, which fulfill the common assumption $\|g - g^\delta\| \leq \delta$. Note that these temperature measurements correspond to values which include the subtraction of $u_c|_{(0,1) \times \Gamma_1}$.

To solve an inverse problem via iterated soft-shrinkage, we need also the adjoint operator K^*. The exact definition of K^* depends crucially on the considered function spaces and has to be investigated carefully. These investigations were done in [5]. It was shown that the adjoint operator can be defined via an adjoint boundary value problem

$$
\begin{aligned}
\frac{\partial v}{\partial t} + \Delta v &= 0 && \text{in } (0,1) \times \Omega \\
v &= 0 && \text{on } (0,1) \times \Gamma_0 \\
-\frac{\partial v}{\partial \nu} &= k && \text{on } (0,1) \times \Gamma_1 \\
v(1, \cdot) &= 0 && \text{in } \Omega.
\end{aligned}
\tag{4.3}
$$

It can be justified that the operator

$$
M : k \longmapsto \frac{\partial v}{\partial \nu}\big|_{(0,1) \times \Gamma_0},
$$

where v denotes the solution of (4.3), defines the adjoint operator of K. See the following remark and the argumentation in [5] for details.

Remark 4.1.1 *Since we deal with noisy measurement data, it is reasonable to consider K as an operator mapping between $L_2(0, 1; L_2(\Gamma_0))$ and $L_2(0, 1; L_2(\Gamma_1))$. However, it turns out that this operator is in general unbounded, whenever the dimension of Ω is greater than 1.*

In [5] the authors introduced spaces

$$
H^{r,s}((0,1) \times S) = H^r(0, 1; L_2(S)) \cap L_2(0, 1; H^s(S))
$$

with associated norm $\|w\|_{H^{r,s}((0,1) \times S)} = \|w\|_{H^r(0,1;L_2(S))} + \|w\|_{L_2(0,1;H^s(S))}$ which are slightly smaller than the L_2-spaces mentioned above. Further, they showed that the operator

$$
K : \{f \in H^{3/4,3/2}((0,1) \times \Gamma_0) \mid f(0, \cdot) = 0\} \to H^{3/4,3/2}((0,1) \times \Gamma_1))
$$

is bounded. Moreover, it was proven that

$$
M : H^{1/4,1/2}((0,1) \times \Gamma_1) \to H^{1/4,1/2}((0,1) \times \Gamma_0)
$$

defines a bounded operator as well and it holds

$$
\langle Kf, k \rangle_{L_2(0,1;L_2(\Gamma_0))} = \langle f, Mg \rangle_{L_2(0,1;L_2(\Gamma_1))}
$$

for all $f \in D(K) \subset L_2(0, 1; L_2(\Gamma_0))$ and $k \in D(M) \subset L_2(0, 1; L_2(\Gamma_1))$. This eventually justifies that the operators are adjoint.

This remark concludes our theoretical considerations of the inverse problem we treat in the current chapter, we now turn to the numerical realization described in the upcoming section.

4.2 Numerical implementation of the reconstruction algorithm

In the current section, how the inverse heat conduction problem has been discretized and how the reconstructions have been computed, are described.

To apply the theory from Chapter 2, it is essential to be provided with an efficient adaptive solver to evaluate the operators K and K^*. For the presented example the finite element toolbox ALBERTA2 was chosen, which is based on the programming language C, see [46] for further information. This toolbox was used, because it provides some features which seemed to be helpful to solve the considered boundary value problems. Firstly, it is relatively easy to do adaptive calculations in time and space dimension and secondly it is possible to define submeshes for extracting parts of the solution at the boundaries. Since it is necessary to extract parts of the solution of the forward problem and to stick them into the adjoint problem as boundary values and vice versa, this feature seemed to be essential.

All the calculations were done using piecewise linear finite elements. This means especially that the degrees of freedom of the approximate solution are located at the nodes of the underlying adaptive mesh. The meshes were created as follows. First the ring-shaped domain Ω was equipped with a fixed macro triangulation stored in an input data file. At the beginning of the solution procedure of the PDEs the macro mesh was refined globally. Afterwards the meshes had been refined and coarsened according to an adaptive strategy, which ensured that the error due to space discretization was distributed equally on the mesh elements. In addition the timesteps had been chosen in such a way that the approximate solution u_l at a fixed time t_l satisfied some error tolerance, i. e.

$$\|u_l - u(t_l)\|_{L_2(\Omega)} \leq h \quad \text{for all } l \in \{0, \cdots, L\},$$

where the $u(t_l)$ denote the exact solutions at these points and L is the index of the timestep corresponding to the endtime. Since the time interval was chosen as $(0, 1)$, this justifies also that the whole approximate solution $u_h = (u_0, u_1, \cdots, u_L)$ satisfies this error bound, i. e.

$$\|u_h - u\|_{L_2(0,1;L_2(\Omega))} \leq h$$

as well as the restriction to the considered part of the boundary.

All calculations presented in the following were done incorporating artificial data g^δ, which had been created as follows. A temperature distribution had been given as Dirichlet boundary data at the inner boundary and the forward problem was solved via the adaptive finite element solver. Afterwards the solution at the outer boundary was extracted, perturbed with Gaussian noise and stored to serve as additional information for the IHCP.

At this point we are provided with all necessary tools and data to solve the IHCP. The whole reconstruction algorithm can be summarized as follows.

Algorithm 4.2.1

1. Choose an initial temperature distribution f^0 at the inner boundary $(0, 1) \times \Gamma_0$.

2. Choose a precision h and set the iteration index $n = 0$.

3. Solve the forward problem (4.2) with $f = f^n$ within a precision of h.

4. Restrict the solution u to the outer boundary $(0, 1) \times \Gamma_1$.

5. Compute the residual $u|_{(0,1) \times \Gamma_1} - g^\delta$.

6. Solve the adjoint problem (4.3) with $k = u|_{(0,1) \times \Gamma_1} - g^\delta$ within a precision of h.

7. Compute the normal derivative of the solution v of the adjoint problem on the inner boundary, i. e. $\frac{\partial v}{\partial \nu}|_{(0,1) \times \Gamma_0}$.

8. Compute the boundary data f^{n+1} for the forward problem by

$$f^{n+1} = \mathbf{S}_{\alpha \mathbf{w}, p}(f^n - (\tfrac{\partial v}{\partial \nu}|_{(0,1) \times \Gamma_0})).$$

9. Go back to step 3.

Having a detailed look at this algorithm, some difficulties arise which will be discussed in the following: First of all, due to the alternating solution of the forward and adjoint boundary value problems, which both involve boundary data based on the solution of the previous problem, it is necessary to interpolate boundary data. This is because the solution method is adaptive and timesteps as well as meshes are in general different at every iteration step. To assemble the linear system for solving the PDEs, several integrals are computed via quadrature rules. To determine the right boundary value in every necessary quadrature point, the corresponding meshes and mesh elements of the previous solution have to be determined and the boundary values have to be interpolated.

Concerning interpolations, we also work with interpolations of the noisy "measurement" data on the outer boundary. Interpolating this data can be seen as a form of data smoothing, which may reduce the ill-posedness of the problem. Nevertheless, this interpolation is a natural assumption, since real data would typically be measured at finitely many locations at the furnace wall and interpolated thereafter.

Another difficulty arises when we have a look at the shrinkage operators. These operators are defined with respect to orthonormal bases of the space $L_2((0,1) \times \Gamma_0) \cong L_2(0, 1; L_2(\Gamma_0))$ containing the boundary values f^n on the inner boundary Γ_0. In our case the arguments $f^n - (\frac{\partial v}{\partial \nu}|_{(0,1) \times \Gamma_0})$ of the shrinkage operator are given in terms of a finite element basis. To fit in with our theoretical findings in Section 2, it would be necessary to expand the finite element solution into some orthonormal basis and shrink the expansion coefficients. We avoid this transformation to get rid of the high numerical costs.

As we will see in the upcoming section, the algorithm produces reasonable results which can even be compared with our theoretical investigations in Section 2.

For a solution of the same IHCP presented in this thesis based on a combination of soft-shrinkage and a wavelet Galerkin scheme, the interested reader is referred to [5]. This wavelet ansatz might match better with our theoretical investigations, since no basis transformation is necessary, for instance. Nevertheless, in this thesis a finite element approach has been chosen since it is widely used and easier to implement than the wavelet Galerkin approach.

4.3   Discussion of numerical results

In the following, some numerical results related to the IHCP presented above will be discussed. All calculations were done with artificial data corresponding to the following model solutions. One data set corresponds to a hotspot, which evolves and vanishes on the inner boundary of the furnace during the considered time period. This setting might not be as realistic as the second data set corresponding to a hotspot which does not vanish again. See Figure 4.2 for an illustration of both model solutions. In the current chapter the boundary data is always plotted on a time space grid, where the space coordinate is given in radian measure.

The first test case deals with the vanishing hotspot. As a regularization method, iterated soft-shrinkage with $p = 1.1$ was used. The data error was fixed to $\delta = 0.085$ which corresponds to a relative error of 5.5%. Since we want to compare our reconstructions with the minimizer of the Tikhonov functional, which is of course unknown, the iteration was done using a small precision of $h = 10^{-3}$ to produce a reference solution. We consider this solution as the true minimizer x_α^δ and will compare our reconstructions to it. Since the reconstructions depend on the regularization parameter, calculations with two different regularization parameters $\alpha = 1$ and $\alpha = 0.5$ were proceeded. Further, three different precisions for the adaptive scheme were used, namely $h = 5 \cdot 10^{-2}$, $h = 10^{-2}$ and $h = 4 \cdot 10^{-3}$. In case of $\alpha = 1$ Figure 4.3 shows the reconstructions after 200 iterations and Figure 4.4 displays the behavior of the distances between the iterates x^n and the minimizer x_α^δ. It can clearly be seen that the distances decrease in every case to a certain level and stay below some bound. This bound gets smaller the smaller the error tolerance of the adaptive solution method is chosen. This is exactly the expected behavior. To compare the numerical computations with the theoretical results of Chapter 2.1, the theoretical bounds $\varepsilon_{\min}(\alpha, \delta, h)$ defined in Proposition 2.1.6 were calculated. In Table 4.1 the bounds obtained by numerical calculations $\varepsilon_{\min,\text{approx}}$ are compared to the theoretical bounds $\varepsilon_{\min,\text{theo}}$ according to Proposition 2.1.6. Obviously

$h \setminus \alpha$	1		0.5	
	$\varepsilon_{\min,\text{approx}}$	$\varepsilon_{\min,\text{theo}}$	$\varepsilon_{\min,\text{approx}}$	$\varepsilon_{\min,\text{theo}}$
$5 \cdot 10^{-2}$	~ 0.77	~ 422	~ 0.95	~ 404304
10^{-2}	~ 0.34	~ 1.39	~ 0.47	~ 4.98
$4 \cdot 10^{-3}$	~ 0.21	~ 0.45	~ 0.33	~ 1.16

Table 4.1: Values for $\varepsilon_{\min}$ due to numerical calculations and theoretical considerations assuming 5.5% relative data error.

the theoretical values are only close to the values obtained by the numerical scheme, if we assume small precisions. For bigger precisions the theoretical values are extremely large, since theoretical considerations assume a worst case scenario. One might have done some more calculations with small precisions, but due to long calculating times it was decided to abandon this. Concerning calculating times, we get back to this point later.

To have a more realistic example, some calculations assuming a non-vanishing hotspot were done. Again, a reference solution using a precision of $h = 10^{-3}$ was computed and serves as an approximation to the Tikhonov minimizer x_α^δ. For this example the data error was fixed to $\delta = 0.0891$ corresponding to 6% relative data error and the regularization parameter was chosen as $\alpha = 0.3$. In Figure 4.5 the reconstructions are pictured, in Figure 4.6 we see again the decreasing distances between the iterates and the reference solution and in Table 4.2 the calculating times are displayed. As in the first example, we

h	time
$5 \cdot 10^{-2}$	30 minutes
10^{-2}	2 hours
$4 \cdot 10^{-3}$	14 hours
10^{-3}	9 days

Table 4.2: Calculating times for 280 iterations assuming a non-vanishing hotspot and 6% relative data error.

see that it is possible to compute reconstructions, which are well localized at the position of the true solution, even without assuming high precision of the adaptive solver. Nevertheless, to get reconstructions, which recover the shape of the true solution, it is necessary to presume a high precision of the adaptive scheme. What is striking in the second example, is that the reconstruction of the non-vanishing hotspot is zero at the end of the time interval. This is due to the considered partial differential equations. The formulation of the adjoint problem (4.3) requires that its solution is zero at the end of the time interval leading to reconstructions, which are zero at that time as well. In general this should be no problem, since in practice one is more interested in the evolution of the hotspot than in its shape at a specific endtime, which can be chosen quite arbitrary.

Finally, a third example will be presented to justify the regularizing properties of the combination of iterated soft-shrinkage and an adaptive finite element solver. As mentioned at the end of the previous section, the numerical realization of the considered IHCP does not fit exactly into the theoretical framework of Chapter 2.1. There are many additional factors, like additional interpolation errors, which influence the numerical results. Nevertheless, some calculations were done assuming a coupling of the data error δ, the regularization parameter α and the precision of the adaptive solver h according to Theorem 2.1.12.

Initially, an approximation $x_a^\dagger$ to the true solution $x^\dagger$ was computed to compare the reconstructions to it. This was done taking noise free data into account, where noise free means that no noise has been added to the artificially computed temperature profile on the outer boundary. But due to interpolations, the data will never be totally noise free and therefore, even if we assume data without additional noise, we will never be able to reconstruct the true solution. Besides noise free data, a small regularization parameter and a small tolerance $h = 10^{-3}$ was used to compute the reference solution $x_a^\dagger$, which is the best reconstruction we can expect. In what follows, we compare the reconstructions to the true solution $x^\dagger$ as well as to the approximation $x_a^\dagger$.

To measure the error between the reconstructions x^N and a reference solution x, we define the relative error by

$$E_{\mathsf{rel}}(x) = \frac{\|x^N - x\|}{\|x\|} \cdot 100\%.$$

Table 4.3 displays how the reconstructions with decreasing noise levels approach the true and approximated true solution. It can be seen that the errors are decreasing for decreasing noise levels regarding the true and approximate true solution. In addition, we see that by reducing the data error from 0.7% to 0.4% the reconstruction errors reduce much more than in the case of larger data errors. The main reason for this behavior might be the high precision h of order 10^{-3}. To be able to verify the theoretical convergence rates, it might have made sense to do some more calculations with higher precision, but due to long calculating times, this has been abandoned. Nevertheless, we can try to estimate convergence rates for our numerical scheme. Figure 4.7 shows on a logarithmic scale the distances between the reconstructions $x^N = x^{N(\delta)}$ and the true and approximated true solution subject to the noise level. Computing in both cases a line of best fit, convergence rates of 0.22 and 0.40 can be deduced, respectively. Since the convergence rates seem to get better for smaller errors, one may suppose that the theoretical convergence rates of at least 0.5, see Theorem 2.1.16, might be achieved considering higher precisions of the adaptive solver or smaller data errors. Finally, in Figure 4.8 the true solution, approximate true solution and the reconstructions for decreasing noise levels are illustrated. Note that the localization of the hotspot in time and space direction is acceptable even for larger data noise and lower precision of the adaptive scheme, whereas it is neccessary to assume a high precision of the adaptive solver to reconstruct the shape of the true solution.

noise	δ	α	h	$\|x^N - x^\dagger\|$	$E_{rel}(x^\dagger)$	$\|x^N - x_a^\dagger\|$	$E_{rel}(x_a^\dagger)$
5.5 %	0.085	0.63	$4 \cdot 10^{-1}$	1.45	58 %	1.24	52 %
2.9 %	0.046	0.34	$1.2 \cdot 10^{-2}$	1.38	55 %	1.10	46 %
1.4 %	0.021	0.16	$2.6 \cdot 10^{-2}$	1.23	49 %	0.89	37 %
0.7 %	0.011	0.08	$6.4 \cdot 10^{-3}$	1.12	45 %	0.80	34 %
0.4 %	0.006	0.04	$1.6 \cdot 10^{-3}$	0.75	30 %	0.37	15 %

Table 4.3: Convergence of the reconstructions $x^N = x^{N(\delta)}$ to the true solution $x^\dagger$ and approximated true solution $x_a^\dagger$ with $\alpha = 7.4\delta$ and $h = \alpha^2$.

To end with, some remarks have to be made. First, it seems as if some of the theoretical results of Chapter 2.1 could only be observed using precisions of the adaptive scheme which are of order 10^{-3} or less. Due to high calculating times a further investigation for smaller precisions was abandoned. Moreover, the results might be improved using different values for the parameters. Since besides the regularization parameter α and the total precision h of the adaptive scheme also the influence of time and space discretization is adjustable, there might of course be better parameter combinations than the chosen ones. Nevertheless, the presented results show that it is possible to solve a non-trivial inverse problem applying a combination of iterated soft-shrinkage and an adaptive finite element scheme.

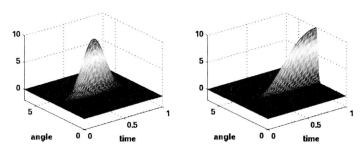

Figure 4.2: True solutions; vanishing hotspot (left) and non-vanishing hotspot (right).

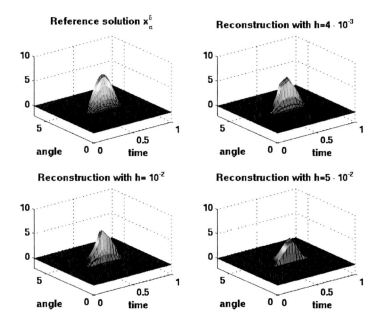

Figure 4.3: Reference solution and reconstructions for $\alpha = 1$.

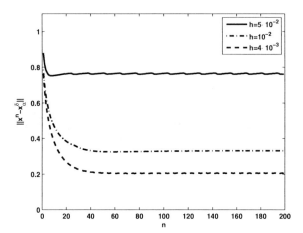

Figure 4.4: Distances between iterates x^n and minimizer x^δ_α for $\alpha = 1$ assuming a vanishing hotspot.

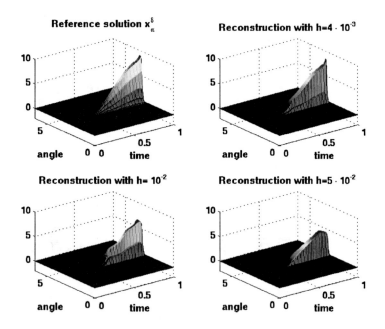

Figure 4.5: Reference solution and reconstructions for a non-vanishing hotspot with $\alpha = 0.3$.

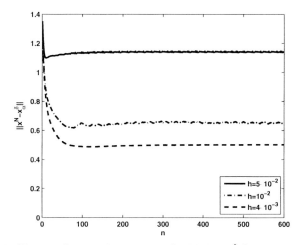

Figure 4.6: Distances between iterates x^n and minimizer x_α^δ for $\alpha = 0.3$ assuming a non-vanishing hotspot.

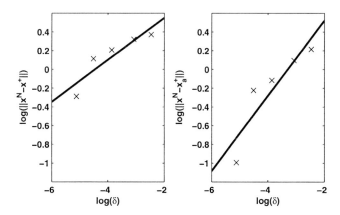

Figure 4.7: Based on the values in Table 4.3 these figures show the lines of best fit which correspond to convergence rates of 0.22 (left) assuming $x^\dagger$ and 0.40 (right) assuming the approximation $x_a^\dagger$ as true solution.

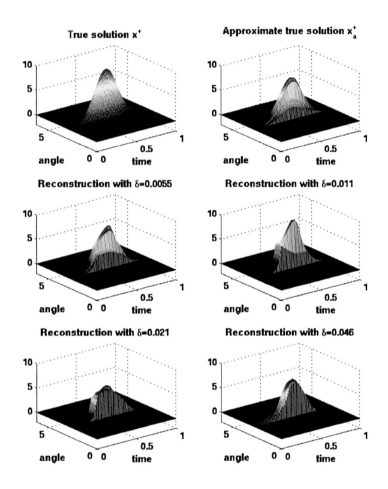

Figure 4.8: True solution, approximate true solution and reconstructions for decreasing noise levels.

Concluding remarks

This thesis contributes to different parts of the theory of inverse and ill-posed problems. The aim of the presented work was to combine the rapidly developing theory of regularization with sparsity constraints with adaptive solution schemes for linear operator equations.

The existing regularization theory typically assumes that the underlying operators can be evaluated exactly. Especially, when adaptive solution schemes are used, this assumption is not valid, but the exact solutions may be computed up to a certain predefined error level. The combination of such adaptive solvers and the iterated soft-shrinkage algorithm proposed by Daubechies, Defrise and De Mol in 2004 [16] is one of the main parts of this thesis. It has been shown that this combination gives a regularization scheme and that the convergence rates carry over from the case including exact operator evaluations. Unfortunately regularizing properties could only be shown in case of weighted ℓ_p-norms with $1 < p \leq 2$, since the contraction property of the involved shrinkage operators is no longer valid for $p = 1$. In addition to adaptive operator evaluations, we studied briefly the case of operator approximations. This means especially that the difference between the operator and its approximation can be estimated in operator norm, which simplifies the analysis. Again, it was possible to prove the regularization property in the case of weighted ℓ_p-norms with $1 < p \leq 2$.

Iterated soft-shrinkage consists basically of alternating Landweber and shrinkage steps and requires quite a lot iteration steps to get proper results. Concerning the high computational costs of adaptive solution schemes, which are for sure lower than those for non-adaptive schemes, it is preferable to do as few iteration steps as possible. Hence, it might be a future goal to combine adaptive schemes with more efficient minimization schemes for Tikhonov-type functionals with sparsity enforcing penalty terms, especially incorporating weighted ℓ_1-norms.

During the investigation of convergence rates for iterated soft-shrinkage, it turned out that all available results seemed to assume an a priori choice of the regularization parameter. This was the motivation to investigate Morozov's discrepancy principle for Tikhonov regularization incorporating non-quadratic penalty terms in Chapter 3. For a special class of penalty terms regularizing properties and convergence rates have been proved. The presented theory covers not only the case of penalty terms consisting of weighted ℓ_p norms, but can be applied to a bigger class of convex functionals.

The combination of Morozov's discrepancy principle, regularization with sparsity constraints and adaptive operator evaluations has been addressed in the last paragraph of the third chapter. However, the presented results are rather technical and give an idea of the complicated coupling of the involved parameters.

Within the third chapter of this thesis, the considered operator equations are formulated in Hilbert spaces. First results, which deal with a formulation in Banach spaces have been worked out by Jin and Zou, see [30].

Within the fourth chapter, an inverse heat conduction problem was investigated. Even though the chosen numerical scheme does not fit exactly into the framework of our investigations within the second chapter, it was shown that a non-trivial inverse and ill-posed problem can be solved via a combination of iterated soft-shrinkage and an adaptive finite element scheme. Also, some theoretical results proved in the second chapter could be justified. Nonetheless, some further investigations should be made in the future. For example, a parameter choice strategy for the regularization parameter has to be involved. Due to the high computational costs the combination of the presented numerical scheme with a discrepancy principle, for instance, was omitted. The regularization parameter was chosen by hand instead, based on visual comparison with the true solution. Since the true solution is usually not available, this cannot be a practical approach. Moreover, the impact of additional interpolation errors due to adaptive grids on the solution, as well as the influence of different weighting parameters to adjust the ratio of accuracy in time and space dimension, might be investigated in more detail in the future.

As mentioned in Chapter 4, a second adaptive scheme has already been combined with iterated soft-shrinkage, namely a wavelet Galerkin method. First results considering the same inverse heat conduction problem as introduced in this thesis have been presented in [5]. To compare both methods and to compute reconstructions incorporating real instead of artificial data will also be future tasks.

Bibliography

[1] H. W. Alt. *Lineare Funktionalanalysis*. Springer, 4th edition, 2002.

[2] G. Aubert and P. Kornprobst. *Mathematical Problems in Image Processing*. Springer, New York, 2002.

[3] T. Bonesky. Morozov's discrepancy principle and Tikhonov-type functionals. *Inverse Problems*, 25(1):Article ID 015015, 2009.

[4] T. Bonesky, K. Bredies, D. A. Lorenz, and P. Maass. A generalized conditional gradient method for nonlinear operator equations with sparsity constraints. *Inverse Problems*, 23(5):2041–2058, 2007.

[5] T. Bonesky, S. Dahlke, P. Maass, and T. Raasch. Adaptive wavelet methods and sparsity reconstruction for inverse heat conduction problems. *Preprint series of the DFG priority program 1324 "Extraktion quantifizierbarer Informationen aus komplexen Systemen"*, 2009.

[6] T. Bonesky and P. Maass. Iterated soft shrinkage with adaptive operator evaluations. *Inverse and Ill-Posed Problems*, 17(4):337–358, 2009.

[7] K. Bredies and D. A. Lorenz. Iterated hard shrinkage for minimization problems with sparsity constraints. *SIAM Journal on Scientific Computing*, 30(2):657–683, 2008.

[8] K. Bredies and D. A. Lorenz. Linear convergence of iterated soft-thresholding. *Journal of Fourier Analysis and Applications*, 14(5-6):813–837, 2008.

[9] K. Bredies, D. A. Lorenz, and P. Maass. A generalized conditional gradient method and its connection to an iterative shrinkage method. *Computational Optimization and Applications*, 42(2):173–193, 2008.

[10] L. M. Bregman. The relaxation method for finding common points of convex sets and its application to the solution of problems in convex programming. *USSR Computational Mathematics and Mathematical Physics*, 7:200–217, 1967.

[11] M. Burger and S. Osher. Convergence rates of convex variational regularization. *Inverse Problems*, 20(5):1411–1420, 2004.

[12] D. Butnariu and E. Resmerita. Bregman distances, totally convex functions and a method for solving operator equations in Banach spaces. *Abstract and Applied Analysis*, 2006:Article ID 84919.

[13] I. Cioranescu. *Geometry of Banach spaces, duality mappings and nonlinear problems*. Kluwer Academic Publishers Group, Dordrecht, 1990.

[14] A. Cohen. *Numerical Analysis of Wavelet Methods*. Elsevier Science B. V., 2003.

[15] P. L. Combettes and V. R. Wajs. Signal recovery by proximal forward-backward splitting. *Multiscale Modeling and Simulation*, 4(4):1168–1200, 2005.

[16] I. Daubechies, M. Defrise, and C. De Mol. An iterative thresholding algorithm for linear inverse problems with a sparsity constraint. *Communications in Pure and Applied Mathematics*, 57(11):1413–1457, 2004.

[17] I. Daubechies, M. Fornasier, and I. Loris. Accelerated projected gradient method for linear inverse problems with sparsity constraints. *Journal of Fourier Analysis and Applications*, 14(5-6):764–792, 2008.

[18] R. A. DeVore. Nonlinear approximation. *Acta Numerica*, 1998:51–150.

[19] I. Ekeland and R. Temam. *Convex Analysis and Variational Problems*. North-Holland, Amsterdam, 1976.

[20] H. W. Engl, M. Hanke, and A. Neubauer. *Regularization of Inverse Problems*, volume 375 of *Mathematics and its Applications*. Kluwer Academic Publishers Group, Dordrecht, 2000.

[21] M. Fornasier. Domain decomposition methods for linear inverse problems with sparsity constraints. *Inverse Problems*, 23(6):2505–2526, 2007.

[22] M. Fornasier and H. Rauhut. Recovery algorithms for vector valued data with joint sparsity constraints. *SIAM Journal on Numerical Analysis*, 46(2):577–613, 2008.

[23] M. Grasmair. Well-posedness and convergence rates for sparse regularization with sublinear l^q penalty term. FSP 092: Joint Research Program of Industrial Geometry, 74, 2008.

[24] M. Grasmair, M. Haltmeier, and O. Scherzer. Sparse regularization with ℓ^q penalty term. *Inverse Problems*, 24(5):1–13, 2008.

[25] R. Griesse and D. A. Lorenz. A semismooth Newton method for Tikhonov functionals with sparsity constraints. *Inverse Problems*, 24(3):Article ID 035007, 2008.

[26] O. Hanner. On the uniform convexity of l_p and ℓ_p. *Arkiv för matematik*, 3(3):239–244, 1956.

[27] F. Hirzebruch and W. Scharlau. *Einführung in die Funktionalanalysis*. Bibliographisches Institut, Mannheim, 1971.

[28] B. Hofmann, B. Kaltenbacher, C. Poeschl, and O. Scherzer. A convergence rates result for Tikhonov regularization in Banach spaces with non-smooth operators. *Inverse Problems*, 23(3):987–1010, 2007.

[29] K. Ito, B. Jin, and J. Zou. A new choice rule for regularization parameters in Tikhonov regularization. Submitted to Numerische Mathematik.

[30] B. Jin and J. Zou. Iterative schemes for Morozov's discrepancy equation in optimizations arising from inverse problems. Submitted to SIAM Journal on Optimization.

[31] J. Lindenstrauß and L. Tzafriri. *Classical Banach spaces II*. Springer, Berlin, 1979.

[32] D. A. Lorenz. Convergence rates and source conditions for Tikhonov regularization with sparsity constraints. *Journal of Inverse and Ill-Posed Problems*, 16(5):463–478, 2008.

[33] A. K. Louis. *Inverse und schlecht gestellte Probleme*. B.G. Teubner, 1989.

[34] P. Maass and A. Rieder. Wavelet-accelerated Tikhonov-Phillips regularisation with applications. In H. W. Engl, A. K. Louis, and W. Rundell, editors, *Inverse Problems in Medical Imaging and Nondestructive Testing*, pages 134–159. Springer-Verlag, 1997.

[35] S. Mallat. *A Wavelet Tour of Signal Processing*. Academic Press, San Diego, CA, USA, 1999.

[36] V. A. Morozov. On the solution of functional equations by the method of regularization. *Soviet Mathematics Doklady*, 7:414–417, 1966.

[37] A. Neubauer. An a posteriori parameter choice for Tikhonov regularization in the presence of modelling error. *Applied Numerical Mathematics*, 4(6):507–519, 1988.

[38] A. Neubauer and O. Scherzer. Finite-dimensional approximation of Tikhonov regularized solutions of non-linear ill-posed problems. *Numerical Functional Analysis and Optimization*, 11(1 & 2):85–99, 1990.

[39] R. Ramlau. Regularization of nonlinear ill-posed operator equations: Methods and applications. *habilitation treatise*, 2004.

[40] R. Ramlau and G. Teschke. Tikhonov replacement functionals for iteratively solving nonlinear operator equations. *Inverse Problems*, 21(5):1571–1592, 2005.

[41] R. Ramlau and G. Teschke. A Tikhonov-based projection iteration for nonlinear ill-posed problems with sparsity constraints. *Numerische Mathematik*, 104(2):177–203, 2006.

[42] R. Ramlau, G. Teschke, and M. Zhariy. A compressive Landweber iteration for solving ill-posed inverse problems. *Inverse Problems*, 24(6):Article ID 065013, 2008.

[43] E. Resmerita. Regularization of ill-posed problems in Banach spaces: convergence rates. *Inverse Problems*, 21(4):1303–1314, 2005.

[44] E. Resmerita and O. Scherzer. Error estimates for non-quadratic regularization and the relation to enhancement. *Inverse Problems*, 22(3):801–814, 2006.

[45] A. Rieder. *Keine Probleme mit Inversen Problemen*. Vieweg, 2003.

[46] A. Schmidt and K. G. Siebert. *Design of Adaptive Finite Element Software.* Springer-Verlag, Berlin Heidelberg, 2005.

[47] F. Schöpfer, A. K. Louis, and T. Schuster. Nonlinear iterative methods for linear ill-posed problems in Banach spaces. *Inverse Problems*, 22(1):311–329, 2006.

[48] T. Schuster, P. Maass, T. Bonesky, K. S. Kazimierski, and F. Schöpfer. Minimization of Tikhonov functionals in Banach spaces. *Abstract and Applied Analysis*, 2008:Article ID 192679.

[49] R. E. Showalter. *Monotone Operators in Banach Space and Nonlinear Partial Differential Equations*, volume 49 of *Mathematical Surveys and Monographs*. American Mathematical Society, 1997.

[50] R. Stevenson. Adaptive solution of operator equations using wavelet frames. *SIAM Journal of Numerical Analysis*, 41(3):1074–1100, 2003.

[51] A. N. Tikhonov and V. Y. Arsenin. *Solution of Ill-posed Problems.* V. H. Winston and Sons, Washington, D.C., 1977.

[52] Z.-B. Xu and G. F. Roach. Characteristic inequalities of uniformly convex and uniformly smooth Banach spaces. *Journal of Mathematical Analysis and Applications*, 157(1):189–210, 1991.

[53] C. A. Zarzer. On Tikhonov regularization with non-convex sparsity constraints. *Inverse Problems*, 25(2):Article ID 025006, 2009.

[54] E. Zeidler. *Nonlinear Functional Analysis and its Applications III.* Springer-Verlag, New York, 1984.

[55] E. Zeidler. *Nonlinear Functional Analysis and its Applications II/A.* Springer-Verlag, New York, 1990.